Green Energy and Technology

More information about this series at http://www.springer.com/series/8059

Andrea Frazzica · Luisa F. Cabeza
Editors

Recent Advancements in Materials and Systems for Thermal Energy Storage

An Introduction to Experimental Characterization Methods

Springer

Editors
Andrea Frazzica
CNR-Istituto di Tecnologie Avanzate per l'Energia "Nicola Giordano"
Messina, Italy

Luisa F. Cabeza
Universitat de Lleida
Lleida, Spain

ISSN 1865-3529 ISSN 1865-3537 (electronic)
Green Energy and Technology
ISBN 978-3-030-07227-8 ISBN 978-3-319-96640-3 (eBook)
https://doi.org/10.1007/978-3-319-96640-3

This Springer imprint is published by the registered company Springer Nature Switzerland AG
The registered company address is: Gewerbestrasse 11, 6330 Cham, Switzerland

Preface

The present book focuses on the experimental methods for the characterisation of innovative thermal energy storage (TES) technologies, comprising high temperature sensible TES, latent TES and thermochemical TES. It provides useful information about the currently employed characterisation methods at lab level, starting from the material up to the systems, trying to elucidate the development level reached by each investigated technology.

It is organised in three main parts, the first one dealing with the analysis of the innovative TES technologies presented in the book; the second one focusing on the experimental characterisation methods for materials and components for TES; the third one dedicated to the experimental characterisation methods for TES systems at lab level.

Part I is subdivided into five chapters. Chapter "Sensible Thermal Energy Storage at High Temperatures" represents a brief introduction to the high temperature sensible TES technology, especially focusing on the most recent achieved advancements. Chapter "Latent Thermal Energy Storage" reports a concise introduction about the latent TES, focusing on the existing phase change materials (PCMs), reporting the main feature of each PCM class along with their pros and cons. Chapter "Thermal Energy Storage with Chemical Reactions" introduces the thermochemical TES technology through chemical reactions. The working principle and the most interesting materials are discussed along with possible applications in different sectors (i.e. automotive, concentrate solar plants and industrial). Furthermore, innovative materials currently under preliminary investigations, such as perovskites, are also introduced. Chapter "Sorption Thermal Energy Storage" focuses on the sorption TES technology, belonging to the wider class of thermochemical TES. Different TES configurations, namely, closed and open cycle, are discussed, and a deep analysis of existing working pairs for sorption TES is reported. Particularly, the main features of liquid absorption, solid adsorption and composite sorbents are discussed and compared. Chapter "Materials Selection for Thermal Energy Storage Applications—Case Studies" describes an innovative methodology for materials selection in TES applications. It represents a useful approach for both researchers and planners, since it allows working on a wide database of TES materials identifying the most promising for the given applications, once properly selected boundary conditions and constraints. The procedure on how to identify and select the constraints is carefully depicted. Finally, some examples applied for storage in buildings and for PCM at high temperature are reported to demonstrate the powerfulness of this method.

Part II is subdivided into five chapters as well. This part mainly focuses on characterization protocols for materials and components for TES. In Chapter "Characterization of Materials for Sensible Thermal Energy Storage at High Temperature", a deep analysis of experimental characterization methods for high temperature sensible TES materials is reported. Thermal characterization techniques are detailed, especially to evaluate crucial parameters such as specific heat and thermal conductivity. Furthermore, also mechanical characterizations are defined, since mechanical resistance represents an important parameter especially for applications in high temperature TES. Chapter "Experimental Methods for the Characterization of Materials for Latent Thermal Energy Storage" reports about experimental methods for the characterization of PCMs for TES applications. Lab methodologies to evaluate main parameters, e.g. latent heat, melting temperature, specific heat, thermal conductivity and cycling stability, are discussed in detail. Subsequently, particular attention is given to novel dynamic experimental characterization of innovative composite PCMs, specifically meant for building applications. Chapter "Experimental Methods for the Characterization of Materials for Thermal Energy Storage with Chemical Reactions" investigates characterization methods for thermochemical TES by means of chemical reactions. The introduced methodologies mainly focus on thermal characterization, by means of thermogravimetric and lab-scale reactors. These approaches allow evaluating reaction rates and degree of conversion for the investigated compounds, which in turn means the evaluation of TES density and power. Furthermore, also morphological and structural methodologies, applied for compounds for TES with chemical reactions, are reported. Indeed, these parameters are crucial especially to evaluate reaction rate hindering as well as possible stability issues. Chapter "Experimental Methods for the Characterization of Materials for Sorption Storage" reports experimental methods for the investigation of sorption materials for TES applications. A detailed section is dedicated to the introduction of equations for sorption equilibrium of a sorbate over a solid sorbent. The experimental methods for the sorption equilibrium measurements are then described, and theoretical and experimental approaches to evaluate the sorption enthalpy are defined and applied to typical sorption working pairs for TES. Chapter "Corrosion Characterization in Components for Thermal Energy Storage Applications" deals with the characterization of corrosion issues occurring in components for TES applications. Different experimental approaches for the corrosion monitoring are introduced. Particularly, the gravimetric and the electrochemical concepts are discussed. The application of the proposed methodologies to different classes of TES materials is discussed, focusing on molten salts for high temperature sensible storage and sorption materials and PCMs for low-medium temperature TES. Finally, corrosion mitigation approaches are introduced to increase the reliability of TES systems.

Part III comprises three chapters, dealing with the experimental methodologies for the characterization of TES systems at lab scale. Chapter "Experimental Characterization of Latent Thermal Energy Storage Systems" reports about experimental approaches to test latent TES system at lab scale. Different testing rigs, belonging to different labs, were described in detail differentiating between 'static' and 'dynamic' latent TES. Furthermore, experimental protocols applied for the evaluation of performance, in terms of TES density and heat transfer efficiency, were introduced. Some of the most interesting results were then reported and discussed. Chapter "Experimental Characterization of Sorption Thermal Energy Storage Systems" focuses on the experimental methodologies for the characterization of lab-scale components and systems for sorption TES. In this chapter, systems operating both in open and closed conditions are discussed. As for the previous chapter, also in this case different testing rigs and testing methodologies are introduced and discussed. Furthermore, also experimental approaches for the testing of components such as adsorbers and evaporators are presented. Indeed, their characterization represents a fundamental step in the development of innovative and high efficiency sorption TES systems. Chapter "Definition of Performance Indicators for Thermal Energy Storage" focuses on the identification and definition of performance indicators needed for a comparable evaluation of TES systems. They were divided into technical, economic and life cycle performance indicators. Finally, the identified indicators were applied to two case studies, a latent and a sorption TES tested in the lab according to the procedures depicted in previous chapters.

We consider the present book as a valuable support for researchers and developing teams in the field of innovative TES systems, to give a clear vision about the current development stage and the experimental approaches developed so far, promoting the discussion on innovative approaches to be applied in the future.

Messina, Italy Andrea Frazzica
Lleida, Spain Luisa F. Cabeza

Acknowledgements

For Chapters "Sensible Thermal Energy Storage at High Temperatures", "Latent Thermal Energy Storage", "Materials Selection for Thermal Energy Storage Applications—Case Studies"–"Experimental Methods for the Characterization of Materials for Latent Thermal Energy Storage", "Corrosion Characterization in Components for Thermal Energy Storage Applications", "Experimental Characterization of Latent Thermal Energy Storage Systems" and "Definition of Performance Indicators for Thermal Energy Storage":

This work was partially funded by the Ministerio de Economía y Competitividad de España (ENE2015-64117-C5-1-R (MINECO/FEDER)).

The authors at the University of Lleida would like to thank the Catalan Government for the quality accreditation given to their research group (2017 SGR 1537). GREA is a certified agent TECNIO in the category of technology developers from the Government of Catalonia.

For Chapters "Materials Selection for Thermal Energy Storage Applications—Case Studies", "Experimental Methods for the Characterization of Materials for Latent Thermal Energy Storage" and "Experimental Characterization of Latent Thermal Energy Storage Systems":

The work partially funded by the Spanish government (ENE2015-64117-C5-2-R).

For Chapters "Materials Selection for Thermal Energy Storage Applications—Case Studies" and "Experimental Methods for the Characterization of Materials for Latent Thermal Energy Storage":

The authors at the DIOPMA would like to thank the Catalan Government for the quality accreditation given to their research group (2017 SGR 118). DIOPMA is a certified agent TECNIO in the category of technology developer from the Government of Catalonia.

Dr. Camila Barreneche has received funding from the European Union's Horizon 2020 research and innovation programme under the Marie Sklodowska-Curie grant agreement No 712949.

For Chapter "Materials Selection for Thermal Energy Storage Applications—Case Studies":

The authors would like to thank the European Commission (FP/2007-2013 N° PIRSES-GA-2013-610692 INNOSTORAGE, Horizon 2020 N°657466 INPATH-TES).

For Chapter "Characterization of Materials for Sensible Thermal Energy Storage at High Temperature":

Angel G. Fernández would like to acknowledge the financial support provided by CONICYT/FONDAP 15110019 'Solar Energy Research Center' SERC-Chile.

For Chapter "Experimental Methods for the Characterization of Materials for Latent Thermal Energy Storage":

Dr. A. L. Pisello would like to thank Fondazione Cassa di Risparmio di Perugia for supporting COLO ARTE project.

For Chapter "Definition of Performance Indicators for Thermal Energy Storage":

Jaume Gasia would like to thank the Departament d'Universitats, Recerca i Societat de la Informació de la Generalitat de Catalunya for his research fellowship (2018 FI_B2 00100).

Contents

Contributors

Camila Barreneche Department of Materials Science and Physical Chemistry, Universitat de Barcelona, Barcelona, Spain; School of Chemical Engineering, BCES (Birmingham Centre of Energy Storage), University of Birmingham, Birmingham, UK

Laura Boquera CIRIAF-Interuniversity Research Centre on Pollution and Environment Mauro Felli, Perugia, Italy; GREiA Research Group, INSPIRES Research Centre, Universitat de Lleida, Lleida, Spain

Vincenza Brancato CNR-ITAE, Istituto di Tecnologie Avanzate per l'Energia "Nicola Giordano", Santa Lucia, Messina, Italy

Luisa F. Cabeza GREiA Research Group, INSPIRES Research Centre, Universitat de Lleida, Lleida, Spain

Luigi Calabrese Department of Engineering, Universita degli Studi di Messina, Contrada di Dio Sant'Agata, Messina, Italy

Alejandro Calderón Department of Materials Science and Physical Chemistry, Universitat de Barcelona, Barcelona, Spain

Ana Inés Fernández Department of Materials Science and Physical Chemistry, Universitat de Barcelona, Barcelona, Spain

Ángel G. Fernández Energy Development Center, University of Antofagasta, Antofagasta, Chile

Andrea Frazzica CNR-ITAE, Istituto di Tecnologie Avanzate per l'Energia "Nicola Giordano", Santa Lucia, Messina, Italy

Margalida Fullana GREiA Research Group, INSPIRES Research Centre, Universitat de Lleida, Lleida, Spain; CIRIAF—Interuniversity Research Centre on Pollution and Environment "Mauro Felli", Perugia, Italy

Shigehiko Funayama Graduate Major in Nuclear Engineering, Tokyo Institute of Technology, Tokyo, Japan

Jaume Gasia GREiA Research Group, INSPIRES Research Centre, Universitat de Lleida, Lleida, Spain

Yukitaka Kato Laboratory for Advanced Nuclear Energy, Department of Chemical Science and Engineering, Major in Nuclear Engineering, Institute of Innovative Research, Tokyo Institute of Technology, Meguro-ku, Tokyo, Japan

Emanuela Mastronardo Thermochemical Processes Unit, IMDEA Energy Institute, Mostoles, Madrid, Spain; Department of Materials Science and Engineering, Northwestern University, Evanston, IL, USA

Candida Milone Department of Engineering, University of Messina, Contrada di Dio, Messina, Italy; National Interuniversity Consortium of Materials Science and Technology (INSTM), Florence, Italy

Valeria Palomba CNR-ITAE, Istituto di Tecnologie Avanzate per l'Energia "Nicola Giordano", Santa Lucia, Messina, Italy

Elpida Piperopoulos Department of Engineering, University of Messina, Contrada di Dio, Messina, Italy; National Interuniversity Consortium of Materials Science and Technology (INSTM), Florence, Italy

Anna Laura Pisello CIRIAF—Interuniversity Research Centre on Pollution and Environment "Mauro Felli", Perugia, Italy; Department of Engineering, University of Perugia, Perugia, Italy

Edoardo Proverbio Department of Engineering, Universita degli Studi di Messina, Contrada di Dio Sant'Agata, Messina, Italy

Joaquim Romaní GREiA Research Group, INSPIRES Research Centre, Universitat de Lleida, Lleida, Spain

Mercè Segarra Department of Materials Science and Physical Chemistry, Universitat de Barcelona, Barcelona, Spain

Salvatore Vasta CNR-ITAE, Istituto di Tecnologie Avanzate per l'Energia "Nicola Giordano", Santa Lucia, Messina, Italy

Part I
Thermal Energy Storage Technologies

Sensible Thermal Energy Storage at High Temperatures

Luisa F. Cabeza

Abstract Sensible thermal energy storage is widely used in different technologies such as buildings and solar power plants, but also in solar cooling, solar drying of food products, and solar cooking. This chapter summarizes the materials mostly used in thermal energy storage, with special highlight to high temperatures (>150 °C).

1 Introduction

Sensible energy storage is the storage of heat with the increase of the temperature of a material (solid, liquid, or gaseous) [1]. Similarly one can talk about the storage of cold with the decrease of the temperature of a material. The energy stored can be expressed with [2, 3]:

$$Q = m \cdot C_p \cdot \Delta T \tag{1.1}$$

where Q is the energy stored [kJ/kg], m is the mass [kg], C_p is the specific heat capacity [kJ/kg K], and ΔT is the temperature change [K].

Thermal energy storage (TES) has applications in all temperature ranges, especially sensible energy storage. This chapter (as the whole book) focusses in sensible energy storage at high temperatures (considered above 150 °C). As listed by Gasia et al. [1], applications with TES at high temperatures are:

- Absorption and adsorption refrigeration (−60 to 350°C).
- Exhaust heat recovery in transportation (55 to 800°C).
- Industry (60–260°C).
- Solar cooling (60–250 °C).

L. F. Cabeza (✉)
GREiA Research Group, INSPIRES Research Centre,
Universitat de Lleida, Pere de Cabrera s/n, 25001 Lleida, Spain
e-mail: lcabeza@diei.udl.cat

A. Frazzica and L. F. Cabeza (eds.), *Recent Advancements in Materials and Systems for Thermal Energy Storage*, Green Energy and Technology,
https://doi.org/10.1007/978-3-319-96640-3_1

- Solar power plants (250–565 °C).
- Industrial waste heat recovery (30–1600 °C).

For an optimal performance, the requirements of high-temperature TES were summarized by Gasia et al. [1]:

- Material requirements:
  - Chemical requirements: long-term chemical stability, no chemical decomposition, compatibility with container materials and low reactivity to heat transfer fluids (HTF), no fire and explosion hazard, no toxicity.
  - Physical requirements: high density, low vapour pressure.
  - Thermal requirements: high specific heat, high thermal conductivity.
- System requirements:
  - Economic requirements: abundant and available, large lifetime, and cost effective.
  - Environmental requirements: low manufacturing energy, easy recycling and treatment, low CO_2 footprint and use of by-products, non-polluting.
  - Technological requirements: operation strategy, integration into the facility, suitable heat transfer between the HTF and the storage medium.

2 Materials

Any material can be used for sensible energy storage, but the most studied or used ones are summarized here [2, 4]:

- Water: It is the most commonly used energy storage material at low temperatures. At high temperature storage is used as steam storage (Ruth system). Its main advantages are its availability, the possibility to be used both as storage materials and as HTF, its high specific heat, non-toxicity, and cost.
- Thermal oils: Although usually used as HTF in solar power plants, they can also be used as storage material.
- Molten salts: These are commercially used in solar power plants and in nuclear plants. Their advantages are high volumetric heat capacity, high boiling point, very high thermal stability, and low vapour pressure. They are also cheap, easily available, non-toxic, and non-flammable. Commercial molten salts are the so-called *solar salt* (60wt% $NaNO_3$ and 40wt% KNO_3), but a lot of research is done for other molten salts such as pure $NaNO_3$, pure KNO_3, HITEC (ternary salt mixture 53% KNO_3, 7% $NaNO_3$, and 40% $NaNO_2$).
- Liquid metals: Aluminium, copper, and lead can be used in sensible storage [5]. Their advantages are high thermal conductivity and high density, but usually the cost is quite high compared to other materials used for TES.

- Earth materials: Rocks, sands, gravel, and others can be used in high temperature sensible heat storage, usually as filler material in thermocline storage tanks with thermal oil as HTF/storage material. They are also used for packed bed fillers (stones, pebbles, and small gravels) and in fluidized bed systems (small granule size sand). Earth materials are cheap, easily available, non-toxic, and non-flammable. More recently, fine sand, ceramics and other are used in solar power plants as direct absorber solar radiation material, as HTF, and as storage material (the three functions with the same material) [6].
- Concrete: Concrete can be used as TES material in solar power plants. Its benefits are low cost, easy construction, good mechanical properties, non-toxic, and non-flammable. One further advantage is that it does not need a container to hold it, which means further costs reduction. Further details can be found in Ndiaye et al. [7].

3 Main Applications

One of the main applications of sensible thermal energy storage at high temperature is in solar power plants (also known as concentrate solar plants—CSP) [8, 9]. Commercial sensible TES is carried out with molten salts, also known as solar salt (60wt% $NaNO_3$ and 60wt% KNO_3). Solar salt is relatively cheap and has a good maximum operating temperature, but requires quite expensive anti-freezing systems, increasing the CAPEX and OPEX of the CSP plants. Other salt systems have been developed and tested, such as Hitec (40% $NaNO_2$, 7% $NaNO_3$ and 53% KNO_3), Hitec XL (48% $Na(NO_3)_2$, 7% $NaNO_3$ and 45% KNO_3), ternary nitrate salts with lithium, and more recently carbonates and chlorides [10].

In order to reduce the cost of the TES system in a CSP plant, researchers propose to substitute the two-tank molten salts configuration by a single tank thermocline system [12]. In this case, a single storage tank is used with strong stratification. To enhance such stratification, a filler material is used. Literature shows that this concept can save up to 35% of investment costs in a parabolic trough solar plant using thermal oil. The solid materials that can be used for this application are metals, sand, bricks, rocks, concrete, and ceramic.

Concrete and castable ceramics were intensively studied for sensible TES in CSP plants [9, 11], since their properties and cost provide good expectations for a good performance and cost. Several additives (such as polypropylene fibres and graphite) are included in the concrete to improve its performance at high temperature. Wastes and by-product materials [13] (asbestos containing wastes, bischofite, slags, etc.) were also studied for such application.

A more novel storage system for CSP is solid particles storage [14]. The first main advantage of this storage system is that solid particles can be used as storage medium as well as heat transfer fluid (HTF) in the same plant. From the material perspective, solid particles systems achieve high performance at high temperature and low cost. Solid particles have three main advantages compared to molten salts:

- Solid particles are chemically inert and stable beyond 1100 °C.
- Solid particles are able to store energy over a higher temperature range.
- Solid particles are expected to have lower costs.

Solid particles TES in CSP is still under research since its use means developing not only the new storage system, but a whole CSP plant design using solid particles as storage material and HTF. Advances in the receiver design, in the particle materials to increase solar absorptivity and durability, and in the balance-of-plant are needed to achieve commercialization of this system.

Another important application of sensible TES is in buildings, but here low temperatures are used. Sensible storage in buildings is mainly implemented with water, but also other materials can be used (common ceramics such as cement and concrete, some natural stones like marble, granite, clay and sandstone, and polymers) [15]. TES can be implemented as a passive technology to provide thermal comfort with the minimum use of energy coming from a HVAC system, or using solar walls. On the other hand, TES can be implemented using active systems in the core of the building (floor or walls), in solar facades, or in water tanks. The same concepts are presented as a good option for zero energy buildings [16, 17].

Sensible TES can also be applied in other applications such as solar cooling [18] (main materials used are water, pressurized water, and thermal oil); solar drying of agriculture and food products [19] (materials used are rocks, brick, concrete, water, oils, and organic liquids such as ethanol); and solar cookers [20] (with materials such as coconut oil, used oil, cast iron, granite, thermal oil, and oil with pebbles).

References

1. Gasia J, Miró L, Cabeza LF (2017) Review on system and materials requirements for high temperature thermal energy storage. Part 1: general requirements. Renew Sustain Energy Rev 75:1320–1338
2. Alva G, Lin Y, Fang G (2018) An overview of thermal energy storage systems. Energy 144:341–378
3. Sarbu I, Sebarchievici C (2018) A comprehensive review of thermal energy storage. Sustainability 10:32
4. Zhang H, Baeyens J, Cáceres G, Degrève J, Lv Y (2016) Thermal energy storage: recent developments and practical aspects. Prog Energy Combust Sci 53:1–40
5. Dinker A, Agarwal M, Agarwal GD (2017) Heat storage materials, geometry and applications: a review. J Energy Inst 90:1–11
6. Calderón A, Palacios A, Barreneche C, Segarra M, Prieto C, Rodriguez-Sanchez A, Fernández AI (2018) High temperature systems using solid particles as TES and HTF material: a review. Appl Energy 213:100–111
7. Ndiaye K, Ginestet S, Cyr M (2018) Thermal energy storage based on cementitious materials: a review. AIMS Energy 6:97–120

8. Liu M, Steven Tay NH, Bell S, Belusko M, Jacob R, Will G, Saman W, Bruno F (2016) Review on concentrating solar power plants and new developments in high temperature thermal energy storage technologies. Renew Sustain Energy Rev 53:1411–1432
9. Pelay U, Luo L, Fan Y, Stitou D, Rood M (2017) Thermal energy storage systems for concentrated solar power plants. Renew Sustain Energy Rev 79:82–100
10. Pfleger N, Bauer T, Martin C, Eck M, Wörner A (2015) Thermal energy storage—overview and specific insight into nitrate salts for sensible and latent heat storage. Beilstein J Nanotechnol 6:1487–1497
11. Ndiaye K, Ginestet S, Cyr M (2018) Thermal energy storage based on cementitious materials: a review. AIMS Energy 6:97–120
12. Grirate H, Agalit H, Zari N, Elmchaouri A, Molina S, Couturier R (2016) Experimental and numerical investigation of potential filler materials for thermal oil thermocline storage. Sol Energy 131:260–274
13. Gutierrez A, Miró L, Gil A, Rodríguez-Aseguinolaza J, Barreneche C, Calvet N, Py X, Fernández AI, Grágeda M, Ushak S, Cabeza LF (2016) Advances in the valorization of waste and by-product materials as thermal energy storage (TES) materials. Renew Sustain Energy Rev 59:763–783
14. Calderon A, Palacios A, Barreneche C, Segarra M, Prieto C, Rodriguez-Sanchez A, Fernández AI (2018) High temperature systems using solid particles as TES and HTF material: a review. Appl Energy 213:100–111
15. de Gracia A, Cabeza LF (2015) Phase change materials and thermal energy storage for buildings. Energy Build 103:414–419
16. Lizana J, Chacartegui R, Barrios-Padura A, Valverde JM (2017) Advances in thermal energy storage materials and their applications towards zero energy buildings: a critical review. Appl Energy 203:219–239
17. Lizana J, Chacartegui R, Barrios-Padura A, Ortiz C (2018) Advanced low-carbon energy measures based on thermal energy storage in buildings: a review. Renew Sustain Energy Rev 82:3705–3749
18. Pintaldi S, Perfumo C, Sethuvenkatraman S, White S, Rosengarten G (2015) A review of thermal energy storage technologies and control approaches for solar cooling. Renew Sustain Energy Rev 41:975–995
19. Kant K, Shukla A, Sharma A, Kumar A, Jain A (2016) Thermal energy storage based solar drying systems: a review. Innovative Food Sci Emerg Technol 34:86–99
20. Nkhonjera L, Bello-Ochende T, John G, King'ondu CK (2017) A review of thermal energy storage designs, heat storage materials and cooking performance of solar cookers with heat storage. Renew Sustain Energy Rev 75:157–167

Latent Thermal Energy Storage

Luisa F. Cabeza

Abstract In the present chapter, an introduction about latent TES technology is reported. Requirements and different types of phase change materials (PCM) are discussed. Finally, a short introduction to PCM composites is presented.

1 Introduction

When energy is stored with the use of the phase change of a material, latent thermal energy storage (also called latent heat storage) is the technology followed [1, 2]. In practical applications, mainly the phase change solid–liquid is used, although the phase change solid–solid can also be of interest. In this technology, when a material is melted, heat is stored at a constant temperature, this is the charge process. The discharge will happen when the material is solidified, and will also happen at constant temperature. It should always be remembered that before and after the phase change, the material will be heated or cooled, storing or releasing energy sensibly (Fig. 1). Moreover, one of the oldest thermal energy storage applications is the use of ice to store cold from winter to summer [3] and the newest one is the use of PCM in high temperature applications [4].

The amount of energy stored is calculated with:

$$Q_{\mathrm{latent}} = \int_{T_1}^{T_{pC}} Cp_{\mathrm{s}} \cdot \mathrm{d}T + \Delta H_{pC} + \int_{T_{pC}}^{T_2} Cp_{\mathrm{l}} \cdot \mathrm{d}T, \qquad (1)$$

where Q_{latent} is the energy stored, Cp_{s} is the heat capacity of the solid, and Cp_{l} is the heat capacity of the liquid.

L. F. Cabeza (✉)
GREiA Research Group, INSPIRES Research Centre,
Universitat de Lleida, Pere de Cabrera s/n, 25001 Lleida, Spain
e-mail: lcabeza@diei.udl.cat

A. Frazzica and L. F. Cabeza (eds.), *Recent Advancements in Materials and Systems for Thermal Energy Storage*, Green Energy and Technology,
https://doi.org/10.1007/978-3-319-96640-3_2

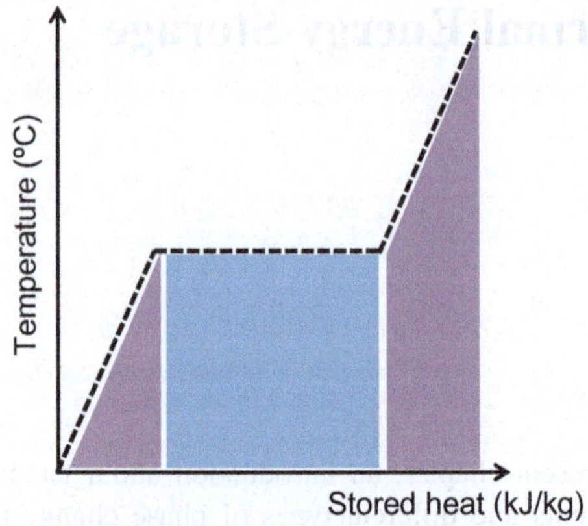

Fig. 1 TES as latent heat for solid–liquid phase change

Table 1 Requirements for latent thermal energy storage [1, 5]

Physical and technical requirements	Thermal requirements	Non-technical requirements
Low density variation and small volume change	Suitable phase change temperature fitted to application	High availability
High energy density	Large phase change enthalpy (ΔH) and specific heat (Cp)	Low cost
Small or no subcooling	High thermal conductivity (except for passive cooling)	Non-toxic
No phase segregation	Reproducible phase change	Non-harmful to the environment
Low vapour temperature	Thermal stability	Recyclable
Chemical and physical stability	Cycling stability	Safety
Compatible with other materials	–	–

As any other technology, several requirements should be considered when selecting the material and designing the components and system. For latent heat storage, those requirements are listed in Table 1. A new methodology for the selection of the right PCM considering some of these requirements was developed by Miro et al. [6].

2 Materials

Already in 1983 Abhat [7] classified the materials used for latent thermal energy storage, those commonly used as phase change materials (PCM). Today it is widely accepted that PCM can be divided between organic and inorganic materials, but more and more, new composites and hybrid materials are being developed. Between organic materials, paraffin and fatty acids are those most used, while between inorganic, salt hydrates and salts are used [8]. Many materials have been studied as PCM and a long list can be found in the literature [8–19]. Advantages and disadvantages of each type of PCM have been reported by many authors [2, 8]:

1. Paraffin: These are the most used commercial PCM. Paraffin is alkanes with different chain length. The longer the alkane the higher the melting temperature. Paraffins show good storage density with respect to mass. As other organic PCMs, they do not show sub-cooling. Their melting temperature is between 0 °C up to around 120 °C. Flammability of paraffin can be a concern in some applications. Finally, paraffin are usually compatible with metals, but not with plastics.
2. Fatty acids: These are also carbon chains, but with one acid radical. Their melting temperature is similar to paraffin and they are stable upon cycling. They show no sub-cooling and low thermal conductivity. Their acid radical can give more corrosivity with metals.
3. Sugar alcohols: They are organic materials with higher melting temperature. Their main disadvantage is that they can oxidise with presence of oxygen, therefore they need to be used in inert atmospheres [20]. Their energy storage density is quite high.
4. Plastics: Plastics such as PEG or HDPE can also be used as PCM. Their main disadvantage is the wider melting enthalpy curve.
5. Water–salt solutions: These materials have a melting temperature below 0 °C, therefore are used for refrigeration and other cooling applications. They usually have good energy storage density. To avoid phase segregation eutectic compositions can be used.
6. Salt hydrates: These are inorganic materials with a melting temperature between 5 and 130 °C. Salt hydrates usually have higher phase change enthalpy than organic PCM, but their main disadvantages are phase segregation, sub-cooling, and corrosion. Their thermal conductivity is similar to water.
7. Salts: These are inorganic materials with melting temperatures above 150 °C that can reach around 800 °C. Their description and advantages and disadvantages are similar to those of salt hydrates, except that salts do no show phase segregation.
8. Mixtures: To improve any of the properties of PCM, mixtures can be used. These can be mixtures of organic materials, of inorganic materials, or of organic–inorganic materials.
9. Metals and metal alloys: Although some pure metals and metal alloys present interesting thermal properties to be used as PCM due to their high storage density and very high thermal conductivity, Fernandez et al. [21] highlighted a lack of understanding on the implications of the metallurgical aspects related to the melting and solidification of these materials under thermal cycling at high temperatures. The main issues to be considered include vapour pressure, sub-cooling, corrosion, segregation, changes in composition and microstructure, changes in thermal properties, and undesired reactions.

3 Composites and Other Material Modifications

A composite is a material composed of several different materials. They are developed to improve a property of a material or to combine properties of different materials. PCM composites can be used to improve the thermal conductivity of the material [22]. PCM composites include:

- Metal foam composite PCM: when the PCM is impregnated into a porous metal foam. As a continuous skeleton structure, the metal foam has high open porosity, high specific strength and stiffness, and high thermal conductivity; but on the other hand, PCM can leak from the composite and corrosion may be an issue.
- Carbon material composite PCM: when the PCM is used together with carbon materials such as expanded graphite, carbon nano-materials, or graphene. The final properties of the composite depend a lot on the type of carbon material used and on the manufacturing process used.
- Form-stable composite PCM [23]: when the PCM is included in a material that prevent the leakage of the PCM. Base materials used are polymers, silica-based materials, perlite, diatomite, clay minerals, etc. The main advantage of these composites is that they do not need to be encapsulated further. The main disadvantage is that the melting enthalpy of the PCM can decrease considerably.

References

1. Mehling H, Cabeza LF (2008) Heat and cold storage with PCM. An up to date introduction into basics and applications. Springer, Berlin
2. Bruno F, Belusko M, Liu M, Tay NHS (2015) Using solid-liquid phase change materials (PCMs) in thermal energy storage systems. In: Cabeza LF (ed) Advances in thermal energy storage systems. Methods and applications. Woodhead Publishing, pp 201–246
3. Paksoy HO (ed) (2007) Thermal energy storage for sustainable energy consumption [electronic resource]: fundamentals, case studies and design. Springer, Dordrecht
4. Cabeza LF, Tay NHS (2018) High-temperature thermal storage systems using phase change materials. Elsevier
5. Barreneche C, Solé A, Martorell I, Martínez M, Cabeza LF, Fernández AI (2014) Thermal energy storage (TES): requirements and constrains for a material-based design. In: Materials Science and Technology Conference and Exhibition 2014, MS and T 2014, vol 3, pp 1651–1658
6. Miró L, Barreneche C, Ferrer G, Solé A, Martorell I, Cazeza LF (2016) Health hazard, cycling and thermal stability as key parameters when selecting a suitable phase change material (PCM). Thermochim Acta 627–629:39–47
7. Abhat A (1983) Low temperature latent heat thermal energy storage: heat storage materials. Sol Energy 30:313–332
8. Zalba B, Marin JM, Cabeza LF, Mehling H (2003) Review on thermal energy storage with phase change: materials, heat transfer analysis and applications. Appl Therm Eng 23:251–283
9. Farid MM, Khudhair AM, Razack SAK, Al-Hallaj S (2004) A review on phase change energy storage: materials and applications. Energy Convers Manage 45(9–10):1597–1615

10. Kenisarin M, Mahkamov K (2007) Solar energy storage using phase change materials. Revew Sustain Energy Rev 11:1913–1965
11. Sharma A, Tyagi V, Chen C, Buddhi D (2009) Review on thermal energy storage with phase change materials and applications. Renew Sustain Energy Rev 13(2):318–345
12. Agyenim F, Hewitt N, Eames P, Smyth M (2010) A review of materials, heat transfer and phase change problem formulation for latent heat thermal energy storage systems (LHTESS). Renew Sustain Energy Rev 14(2):615–628
13. Haillot D, Bauer T, Kröner U, Tamme R (2011) Thermal analysis of phase change materials in the temperature range 120–150 °C. Thermochim Acta 513(1–2):49–59
14. Oró E, de Gracia A, Castell A, Farid M, Cabeza LF (2012) Review on phase change materials (PCMs) for cold thermal energy storage applications. Appl Energy 99:513–533
15. Liu M, Saman W, Bruno F (2012) Review on storage materials and thermal performance enhancement techniques for high temperature phase change thermal storage systems. Revew Sustain Energy Rev 16(4):2118–2132
16. Li G, Hwang Y, Radermacher R, Chun HH (2013) Review of cold storage materials for subzero applications. Energy 51:1–17
17. Brancato V, Frazzica A, Sapienza A, Freni A (2016) Identification and characterization of promising phase change materials for solar cooling applications. Sol Energy Mater Sol Cells 160:225–232
18. Chandel SS, Agarwal T (2017) Review of current state of research on energy storage, toxicity, health hazards and commercialization of phase changing materials. Revew Sustain Energy Rev 67:581–596
19. Du K, Calautit J, Wang Z, Wu Y, Liu H (2018) A review of the applications of phase change materials in cooling, heating and power generation in different temperature ranges. Appl Energy 220:242–273
20. Solé A, Neumann H, Niedermaier S, Martorell I, Schossig P, Cabeza LF (2014) Stability of sugar alcohols as PCM for thermal energy storage. Sol Energy Mater Sol Cells 126:125–134
21. Fernández AI, Barreneche C, Belusko M, Segarra M, Bruno F, Cabeza LF (2017) Considerations for the use of metal alloys as phase change materials for high temperature applications. Sol Energy Mater Sol Cells 171:275–281
22. Zhang P, Xiao X, Ma ZW (2016) A review of the composite phase change materials: Fabrication, characterization, mathematical modeling and application to performance enhancement. Appl Energy 165:472–510
23. Fang G, Tang F, Cao L (2014) Preparation, thermal properties and applications of shape-stabilized thermal energy storage materials. Renew Sustain Energy Rev 40:237–259

10. Kenisarin M, Mahkamov K (2007) Solar energy storage using phase change materials. Renew Sustain Energy Rev 11(9):1913–1965
11. Sharma A, Tyagi V, Chen C, Buddhi D (2009) Review on thermal energy storage with phase change materials and applications. Renew Sustain Energy Rev 13(2):318–345
12. Agyenim F, Hewitt N, Eames P, Smyth M (2010) A review of materials, heat transfer and phase change problem formulation for latent heat thermal energy storage systems (LHTESS). Renew Sustain Energy Rev 14(2):615–628
13. Haillot D, Bauer T, Kröner U, Tamme R (2011) Thermal analysis of phase change materials in the temperature range 120–150 °C. Thermochim Acta 513(1–2):49–59
14. Oró E, de Gracia A, Castell A, Farid MM, Cabeza LF (2012) Review on phase change materials (PCMs) for cold thermal energy storage applications. Appl Energy 99:513–533
15. Liu M, Saman W, Bruno F (2012) Review on storage materials and thermal performance enhancement techniques for high temperature phase change thermal storage systems. Renew Sustain Energy Rev 16(4):2118–2132
16. Li G, Hwang Y, Radermacher R, Chun HH (2013) Review of cold storage materials for subzero applications. Energy 51:1–17
17. Brancato V, Frazzica A, Sapienza A, Freni A (2016) Identification and characterization of promising phase change materials for solar cooling applications. Sol Energy Mater Sol Cells 160:225–232
18. Chandel SS, Agarwal T (2017) Review of current state of research on energy storage, toxicity, health hazards and commercialization of phase changing materials. Renew Sustain Energy Rev 67:581–596
19. Du K, Calautit J, Wang Z, Wu Y, Liu H (2018) A review of the applications of phase change materials in cooling, heating and power generation in different temperature ranges. Appl Energy 220:242–273
20. Solé A, Neumann H, Niedermaier S, Martorell I, Schossig P, Cabeza LF (2014) Stability of sugar alcohols as PCM for thermal energy storage. Sol Energy Mater Sol Cells 126:125–134
21. Fernández AI, Barreneche C, Belusko M, Segarra M, Bruno F, Cabeza LF (2017) Considerations for the use of metal alloys as phase change materials for high temperature applications. Sol Energy Mater Sol Cells 171:275–281
22. Zhang P, Xiao X, Ma ZW (2016) A review of the composite phase change materials: fabrication, characterization, mathematical modeling and application to performance enhancement. Appl Energy 165:472–510
23. Fang G, Tang F, Cao L (2014) Preparation, thermal properties and applications of shape-stabilized thermal energy storage materials. Renew Sustain Energy Rev 40:237–259

Thermal Energy Storage with Chemical Reactions

Candida Milone, Yukitaka Kato and Emanuela Mastronardo

Abstract The chapter addresses the main issues dealing with four types of reversible processes, such as dehydration of salt hydrates and hydroxides, thermal decomposition of oxides and perovskites for thermal energy storage as example of thermochemical processes covering a broad range of temperature heat sources.

1 Introduction

Thermal energy storage (TES) in the form of chemical energy, also called termochemical TES, represents a valid alternative to the traditional sensible and latent TES due to higher storage density, longer storage time with lower thermal dissipation [1].

Thermochemical TES is realized performing a reversible chemical reaction. Based on simple thermodynamic considerations the reaction in one direction will be

C. Milone (✉)
Department of Engineering, University of Messina,
Contrada di Dio 98166, Messina, Italy
e-mail: cmilone@unime.it

C. Milone
National Consortium for Materials Science and Technology,
INSTM, Via G. Giusti, 9, 50121 Florence, Italy

Y. Kato
Department of Chemical Science and Engineering,
Major in Nuclear Engineering, Institute of Innovative Research,
Tokyo Institute of Technology, N1-22-2-12-1 O-okayama,
Meguro-ku, Tokyo 152-8550, Japan

E. Mastronardo
Thermochemical Processes Unit, IMDEA Energy Institute,
Mostoles, Madrid 28935, Spain

E. Mastronardo
Department of Materials Science and Engineering,
Northwestern University, Evanston, IL 60208, USA

A. Frazzica and L. F. Cabeza (eds.), *Recent Advancements in Materials and Systems for Thermal Energy Storage*, Green Energy and Technology,
https://doi.org/10.1007/978-3-319-96640-3_3

endothermic while in the other direction will be exothermic (Eq. 1). A general description of the process is represented in Eq. 1,

$$AB + \Delta H_{reac} \leftrightarrow A + B \quad (1)$$

The compound AB is converted into the initial components A and B through an endothermic reaction. This represents the "*charge stage*" since reaction products, A and B, stores thermal energy in the form of the chemical potential energy and is driven by the thermal power transferred from the surroundings. Separation of the reaction products A and B is strictly required for conservation of stored energy and the storage time is practically decided by the time in which the products are kept separated.

The exothermic recombination of A and B to AB represents the "*discharge stage*" and results in giving stored heat back.

The working principle of the thermochemical TES is also depicted in Fig. 1. It is evident that the system is able to keep thermal energy stored without any loss as long as the two reaction elements are kept separated. This feature is of great importance, indeed it allows considering this technology as one of the most promising for long-term (seasonal) TES.

Among the several reversible chemical processes for storing heat such as gas–gas reactions, liquid–gas reactions, and solid–gas reactions [3] attention will be paid to solid–gas reactions.

Thanks to the highest density of solids, higher storage energy density is achieved that is more compact and easy to handle devices. Moreover, the use of solid facilitates the separation of gaseous phase formed upon the chemical reaction and undergoes to small volume variation during reaction.

Working temperature is a key parameter for the choice of thermochemical TES process hence, the thermochemical material (TCM), which in turn depends on the requirements of the final application. The turning temperature T^* of the reversible chemical reactions, i.e. the temperature at which the standard Gibbs-free energy change of reaction is equal to zero, is the first parameter to look at to determine which chemical reactions might be selected for the given application.

At T^* both the reactants and products exist; the *charge stage* occurs at $T > T^*$ while the *discharge stage* prevails when $T < T^*$. Hence the turning temperature T^* of the reaction should be approximately in the range between the temperature values (T_c and T_d) of the thermal energy sources at which charge and discharge has to be realized ($T_d < T^* < T_c$) [3].

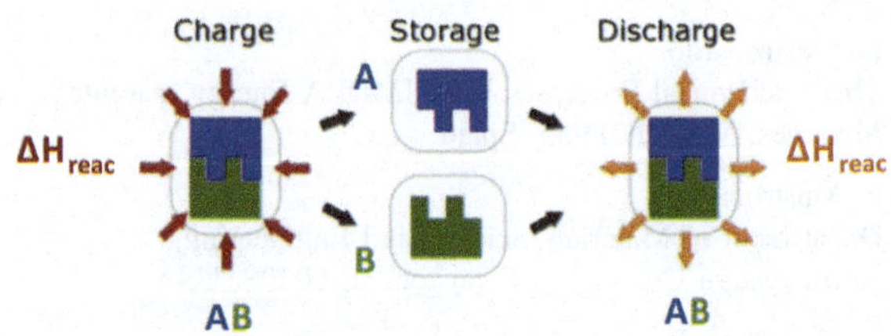

Fig. 1 Description of the thermochemical process TES, adapted from [2]

The ideal thermochemical TES process should also have the following characteristics:

- High reaction heat;
- Good reversibility;
- Fast charging and discharging rates;
- Stable reaction products;
- Non-toxic, non-corrosive, non-flammable and non-explosive reactants and products;
- Large-scale availabilities and abundance, cheap price.

In the following paragraphs recent advancement on TCMs and related reactions are presented.

In particular, the main issues dealing with four types of processes such as dehydration of salt hydrates and hydroxides, thermal decomposition of oxides and perovskites will be discussed as example of TCMs covering a broad range of temperature for TES applications, from low to high temperature in the order, will be discussed.

2 Dehydration of Salt Hydrates

Storage and release of low temperature heat by using salt hydrates rises from the reversible reaction below reported:

$$\text{Salt}\, x\text{H}_2\text{O}_{(s)} + \text{Heat} \leftrightarrow \text{Salt}_{(s)} + x\text{H}_2\text{O}_{(g)} \quad (2)$$

The de/hydration reaction can occur in different steps at different desorption temperatures and intermediate hydrates phases can appear in the system.

Cuypers et al. [4], on the base of theoretically energy density (reachable in an ideal process), price and workability identify $MgCl_2$, Na_2S, $CaCl_2$ and $MgSO_4$ as the most promising candidates for thermochemical TES. The theoretical energy density ranges between 1.85 and 3.17 GJ/m^3 [4] with Na_2S showing the highest theoretical value (3.17 GJ/m^3 in the complete dehydration of $Na_2S \cdot 9H_2O$). As expected, under operating conditions, a lowering of energy density with respect to the theoretical values was found for most of the above reported salts, however the highest value was still achieved for Na_2S [2]. The main disadvantages in the use of this salt is the formation of toxic and corrosive H_2S gas during hydration [4]. Similarly, $MgCl_2$ use is strongly limited also by the HCl formation at temperature lower than 203 °C [5].

Deliquescence (overhydration) and/or melting point lower than hydration temperature, i.e. low stability of materials under operating conditions, are the main drawbacks encountered in the use of $CaCl_2$ [6], $MgSO_4$ [7] and Na_2SO_4 [4].

Aiming at improving the stability of salt hydrates, the main strategies adopted are encapsulation of active phase and impregnation in porous inactive materials to realize the composite sorbent concept [8–11]. This class of materials is often included in the sorption class, which will be discussed in detail in the following chapter. Nevertheless, some of the most recent outcomes on the development of sorbent composites are discussed below.

Porsen et al. [9] studied the de/hydration behaviour of $MgSO_4$ deposited over vitreous host materials having a broad range of pore size (7 nm–1.7 μm). $MgSO_4$ content in the composites ranged between 0.80 and 1.89 mmol of anhydrous $MgSO_4$ per gram of host material. It was found that upon dehydration of $MgSO_4 \cdot 7H_2O$ composites at 40 °C the loss of water molecules decreased with respect to not-hosted sample but for matrix having the higher pore size [9], as a result of the kinetic hindrance of the dehydration reaction. For all the samples, the dehydration at higher temperature (130 °C) led to the formation of a mixture of hydrates.

Water uptake at 30 °C and 85% RH increased with decreasing pore size and in most cases exceeded the stoichiometric water content for the formation of $MgSO_4 \cdot 7H_2O$, the thermodynamically most stable solid phase at 30 °C. Lowering the pore sizes of hosting materials decreased the amount of water of crystallization and increased the amount of water condensed into the pores.

Heats of reaction, determined by calorimetry at 85% RH and 30 °C, exceeded 1.82 kJ/g measured for not-hosted active phase, except for vitreous materials having larger pore sizes (1.7 μm). Higher volumetric energy density (500–550 MJ/m^3) values were achieved for composites having the lower host matrix pore size, mainly due to the contribution of water vapour capillary condensation into pores.

Kallenberger et al. [8] investigated $MgSO_4$/polymer composites having a salt content of 80 wt%. $MgSO_4 \cdot 1.25H_2O$ was prepared by treatment of hydromagnesite tetrahydrate ($(Mg_5CO_3)_4(OH)_2 \cdot 4H_2O$) previously encapsulated into a phenol-formaldehyde resin with sulphuric acid. Results showed that in order to create porous coating resin the tripolymer block Pluronic F127 have to be used.

Hydration, performed at 30 °C and 81% RH, led to a mixture of $MgSO_4 \cdot 6H_2O$ and $MgSO_4 \cdot 7H_2O$ and a heat release of 1080 kJ/kg. The volumetric heat release is 930 GJ/m^3. Composites undergone to cyclic experiment (40 cycles) cracked forming a powder like substances and the hydration enthalpy increased as a result of the higher accessibility of the salt within the sulphate composite [8].

Gaeini et al. [11] reported that impregnation of $CaCl_2$ in expanded graphite or vermiculite did not allow to overcome the salt overhydration under operating conditions. Microencapsulation of $CaCl_2$ into ethyl cellulose, instead, improved the stability of the material over multiple cycles. Reaction occurred with a fast kinetics however, the resulting energy density storage was lower [11].

The research activity for novel salt hydrates and/or their combination, having specific properties related to the precise application, are nowadays vivacious [9, 12–17].

Theoretical approach was proposed by Donkers et al. [17] and Deutsch et al. [16] while both theoretical and experimental approach was proposed by N'Tsoukpoe et al. [15].

Donkers et al. [17] collected and analysed the thermodynamic data of a large number of salt hydrate reactions (563 reactions) aiming at evaluating the theoretical possibilities and limitations of salt hydrates as thermochemical materials (TCMs) for seasonal TES in the built environment (1 GJ/m^3 on system level). Filtering data by hydration reaction capacity higher than 1.3 GJ/m^3, a hydration temperature of 50 °C or higher, a dehydration temperature below 120 °C a shortlist of 25 TCM hydrate reactions was identified. The shortlist was then analysed by considering price, chemical stability, reaction kinetics and safety for domestic environment. At the end of the process K_2CO_3 was identified as the most promising candidate for open or closed systems, but it had a low energy density [17]. It is worthy to note that, based on the review of 563 hydrate reactions, it was concluded that no ideal salt exists under the considered boundary conditions.

N'Tsoukpoe et al. [15], upon fixing as target application a storage unit that should cover both heating and domestic hot water heat demand, with 60 °C as minimum discharging temperature, performed a systematic evaluation of the suitability of 125 salt hydrates in a three-step approach. Preliminary selection was done considering toxicity and risk of explosion; in the second and third steps, the authors implemented a combined approach consisting of theoretical calculations and experimental measurements using thermo-gravimetric Analysis (TGA). The net energy storage density of the material and the thermal efficiency were used to evaluatc the potential of 45 preselected salt hydrates for a low temperature thermochemical TES application.

$SrBr_2{\cdot}6H_2O$ and $LaCl_3{\cdot}7H_2O$ appeared to be the most promising but only from thermodynamic point of view, taking no notice of economical evaluation. Analogously to other authors [17] it was concluded that as long as the classical thermochemical TES processes will be used the thermal efficiency remains low. They agreed on the need to adopt strategies to change some properties of a salt (melting point or the equilibrium curves) proposing addition of a supplementary salt (mixing or synthesis of compounds with mixed cations or anions).

Posern et al. [9] investigated the thermochemical behaviour of $MgSO_4$ and $MgCl_2$ mixture hydrates impregnated in attapulgite (a magnesium aluminium phyllosilicate). The reduction of the deliquescence relative humidity (DRH) of the mixture by the partial substitution of $MgSO_4$ by $MgCl_2$ increased the capacity of condensation and therefore the released heat. The energy density of the composite containing 20 wt% $MgSO_4$ and 80 wt% $MgCl_2$ was 1590 kJ/kg, measured by calorimetric analysis (at 30 °C, 85% RH) with a desorption temperature of 130 °C.

Knoll et al. [14] investigated the de/hydration behaviour and cycle stability of calcium oxalate ($CaC_2O_4{\cdot}H_2O$). The selection of the material came out by the application of an algorithm capable of a systematic database evaluation of potentially suitable reversible reactions [16]. Calcium oxalate showed hydration temperature higher than 100 °C and slightly below the charging temperature. The reversibility of the reaction was guaranteed by the change in water pressure.

Gutierrez et al. [12] pointed out that the strategy of modification of TCMs in order to improve their performance or to reach a determined requirement could affect their cost thus limiting the applicability. To face with economic aspect, they turned the attention towards several wastes and by-products from the non-metallic mining, such as salt hydrates and double salts, largely available and without any application. In this frame astrakanite ($Na_2SO_4{\cdot}MgSO_4{\cdot}4H_2O$), lithium carnallite ($LiCl{\cdot}MgCl_2{\cdot}7H_2O$) and potassium carnallite ($KCl{\cdot}MgCl_2{\cdot}6H_2O$) were characterized as TES materials. The costs of these materials are close to zero. The results showed that astrakanite and potassium carnallite have potential to be applied as TCM at low-medium temperature (<300 °C). In addition, a dehydrated product obtained from astrakanite showed potential to be applied as phase change material (PCM) from 550 to 750 °C. On the contrary, lithium carnallite did not show potential to be applied as TES material due to its low thermal stability, presenting partial decomposition below 200 °C.

3 Dehydration of Hydroxides

Metal oxide hydration and dehydration shown in Eq. 3 can be used for thermochemical energy storage. M, MO_m and $M(OH)_n$ depict metal, metal oxide and metal hydroxide, respectively.

$$MO_m(s) + (n - m)H_2O(g) = M(OH)_n(s) \tag{3}$$

Relationship between $1/T$ and ln $K(=P_{H_2O})$ is linear at a range in which changes of reaction enthalpy and entropy are negligible, and is called as a reaction equilibrium line. Figure 2 shows reaction equilibrium lines of metal oxide/water reaction systems [18]. The temperature at reaction (vapour) pressure of 101.3 kPa is T_{turn}. Calcium oxide (CaO)/H_2O is the most popular reactant [19]. It is recognized that nickel oxide (NiO) and cobalt oxide have potential to be reaction candidates for TES at less than 200 °C, and strontium oxide (SrO) and barium oxide (BaO) could be candidates for the storage at over 700 °C. Calcium sulphate ($CaSO_4$)/H_2O and metal sulphate [20], calcium chloride ($CaCl_2$)/H_2O and metal chloride [21], strontium bromide ($SrBr_2$)/H_2O and metal bromide [22] are also useful systems for TES at around 100 °C.

Magnesium oxide/water (MgO/H_2O) TES is studied for utilization of exhaust heat of medium temperature at around 200–400 °C from an internal combustion engine and industrial processes [23, 24], according to Eq. 4.

$$MgO(s) + H_2O\,(g) = Mg(OH)_2(s), \quad \Delta H = 81.0\ \text{kJ mol}^{-1} \tag{4}$$

Material developments of MgO/H_2O system are discussed for enhancement of the TES performance. Lithium chloride-modified magnesium hydroxide ($Mg(OH)_2$) [25] and lithium bromide one [26] show higher TES performance than authentic magnesium hydroxide by water absorption enhancement effect of them on the surface of MgO particle. Although the $Mg(OH)_2$ pellet has high reactivity, thermal conductivity enhancement of material is important for efficient heat exchange.

Fig. 2 Reaction equilibrium lines of metal oxide/water reaction systems [18]

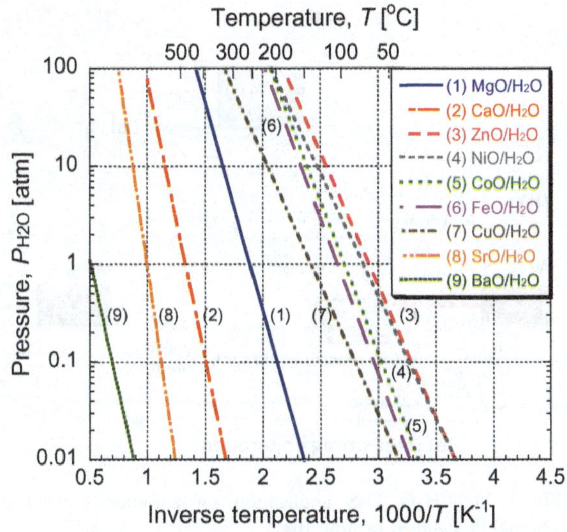

The packed bed of $Mg(OH)_2$ pellets is characterized by a low effective thermal conductivity, around 0.2 W m^{-1} K^{-1}. Expanded graphite (EG) is a good candidate for thermal conductivity enhancement. $Mg(OH)_2$ composite mixed with EG, named as EM, was developed in previous works [27–30]. In comparison with pure-Mg $(OH)_2$, EM has mould-ability property, which means capability of being shaped easily into a specific figure by compaction in a mould. Measurements of thermal conductivity of EM tablets showed higher effective thermal conductivity up to 2.0 W m^{-1} K^{-1}. The mixing mass ratio between $Mg(OH)_2$ and EG of 8:1 showed optimal performances among others by a packed bed reactor experiment [31]. Thermal conductivity enhancement and reactor design of TES material are important issue for practical TES system. Carbon nanotubes and graphene oxide based hybrid materials for MgO/H_2O was also developed for reactivity enhancement by heat transfer enhancement effect of carbon nanotube [32, 33].

MgO/H_2O TES is studied for utilization of exhaust heat of medium temperature at around 200–400 °C from an internal combustion engine. The planned application of the TES on automobile catalyst heating is shown in Fig. 3a. Internal combustion engine in automobile emits exhaust gas containing nitrogen oxide (NO_x), toxic gas, and NO_x should be reduced into nitrogen and oxygen by an on-board catalytic reactor at 300 °C. At a daily start of the engine, the catalyst should be firstly heated up to the active temperature from atmospheric temperature. The heating operation requires additional fuel. MgO/H_2O TCES could assist to supply heat to the reactor by storing surplus heat from exhaust gas under normal running on the day before (Fig. 3a), and outputting the stored heat to the catalyst by hydration (Fig. 3b), then, realize fuel consumption save and CO_2 emission mitigation.

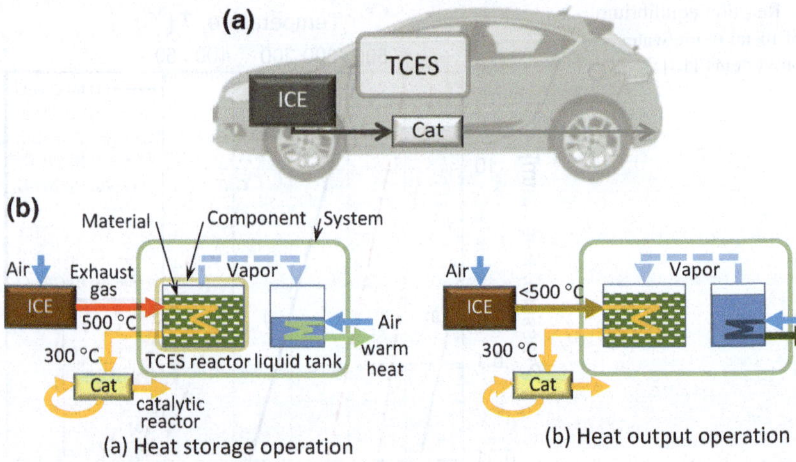

Fig. 3 MgO/H_2O TES application on automobile catalyst heating (**a**) and operation for automobile catalyst heating (**b**)

4 Thermal Decomposition of Metal Oxides

Reducible metal oxides, exhibiting a rich chemistry related to changes of the metal oxidation state, represents a suitable category of compounds for storing heat from high temperature source such heat transfer fluid (HTF) from concentrated solar power (CSP) [34].

The redox reversible reaction is simplified in Eq. 5:

$$MO_{x(s)} + \text{Heat} \leftrightarrow MO_{x-y(s)} + \frac{y}{2}O_{2(g)} \tag{5}$$

Endothermic reduction reaction allows storing thermal energy during on sun conditions while exothermic oxidation reaction releases the stored heat during off sun conditions.

Redox cycles of metal oxides can proceed in an open system at temperature higher than 500 °C, depending on the redox pairs, in air (pO_2 = 0.2 atm) [35, 36]. Such operating conditions allow the use of a hot air stream from a volumetric receiver.

From Fig. 4, where is shown the plot Equilibrium Temperature versus pO_2 for a number of redox pairs [35], it is possible to verify the conditions required for reduction/oxidation in order to take a decision on the more suitable redox pair depending on the temperature of HTF and the application.

In Fig. 5 the specific energy storage of redox pairs versus the atomic molar energy storage is reported [35]. The amount of energy stored in the reduction step is obviously of utmost importance for heat storage application, however fast reaction

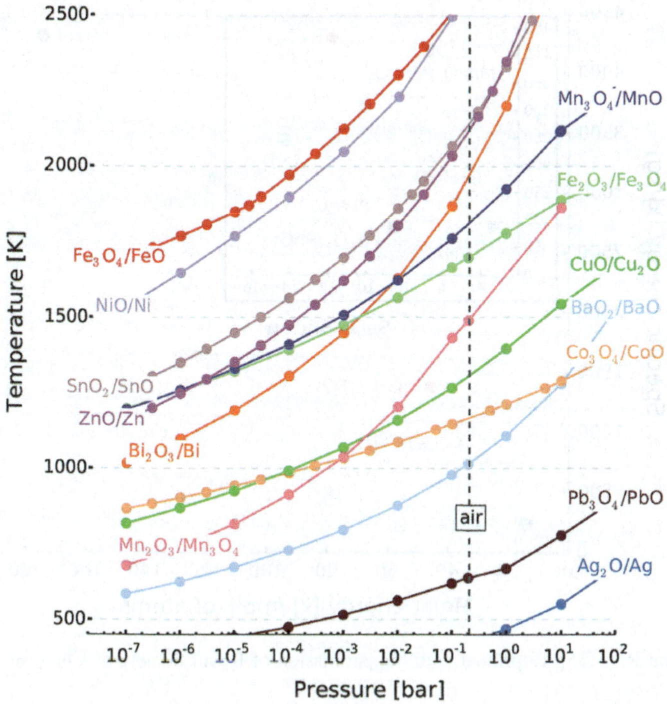

Fig. 4 From Ref. [35] (thermodynamics data are from software Factsage). Reprinted with the permission of Royal Society of Chemistry

kinetics, long-term stability and low materials costs are also required for considering materials for real applications [36].

Among reduction pairs reported in Fig. 3 Co_3O_4/CoO, CuO/Cu_2O, Fe_2O_3/Fe_3O_4 represent a good compromise between specific energy stored and reduction temperature, see Fig. 4. CeO_2 suffers of high reduction temperature and low specific energy storage due to the non-stoichiometric reduction of CeO_2 [35].

Below, a brief overview of the study carried out with pure and mixed metal oxides is reported. Unless otherwise specified redox reduction/oxidation steps has been carried out at $pO_2 = 0.2$ atm.

Co_3O_4/CoO is one of the most investigated redox pair [36–43]. Reduction occurs at 914 °C and the measured enthalpy of reaction is 844 kJ/kg [36, 38]; the material shows a good reversibility. The main drawbacks are the low cyclability due to sintering which leads to a decrease of the reaction rates with increasing numbers of cycles [36, 38], the high material cost and toxicity [38].

Manganese Sesquioxide (Mn_2O_3) is harmful less rather inexpensive material that reduces at 942 °C [36, 38]. However it is characterized by a low reduction enthalpy (230 kJ/kg in the reduction Mn_2O_3 to Mn_3O_4) and scarce reversibility in air [36, 38].

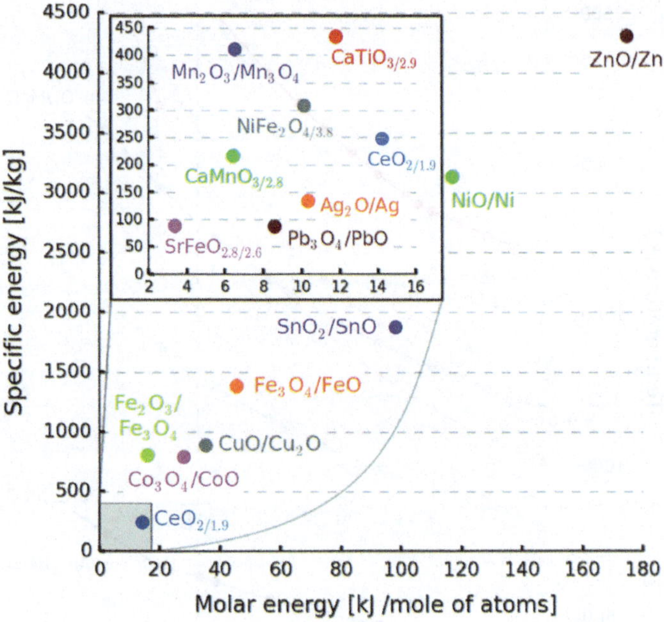

Fig. 5 From Ref. [35]. Reprinted with the permission of Royal Society of Chemistry

Müller et al. [39] reported that redox cycle carried out at $pO_2 = 1$ atm leads to a complete reoxidation of Mn_3O_4 while only 72.8% of Mn_3O_4 is reoxidized in air, under their experimental conditions.

Carrillo et al. [44] observed that Mn_2O_3 primary particle size influenced reoxidation behaviour and the cyclability. The decrease of Mn_2O_3 primary particle size lowered the oxidation temperature, slowered the rate of reaction and influenced the sintering mechanism and then the cyclability. Specifically, larger primary particle size caused a high degree of coarsening thus presenting a more open structure with larger pores, which are more favourable for O_2 diffusion while smaller primary particles size led to more densely packed structures.

Experimental studies on CuO/Cu_2O pair showed reduction enthalpy of 650 kJ/kg; the reaction takes place at 1042 °C [36, 44–47]. The reduction temperature, however, is close to the melting point of copper oxide (1350 K for $pO_2 = 0.21$ atm) resulting in significant grain growth after thermal cycling by Simultaneous Thermal Analysis (STA) [36]. CuO showed the fastest reduction reaction of the pure metal oxides investigated, nevertheless, the reoxidation is slow, mainly due to the pronounced sintering effects of this metal oxide [36]. Conversely, Deutsch et al. [45] did not observe dramatic material sintering when reduction is carried out, by STA, in pure nitrogen although the temperature (950 °C) is close to the melting point of an eutectic CuO/Cu_2O (1091 °C and 35% of CuO).

Alonso et al. [47] studied the performance of CuO reduction in a rotary kiln solar furnace in argon or air. Reduction of CuO in argon atmosphere led to conversions of almost 80%, higher than in air. Although the rotation movement generated the particles grouping in small spheres, the consistence of the material did not change in case of reduction with argon while stronger coalescence of the particles was found in air, which hinder oxide conversion.

Fe_2O_3/Fe_3O_4 redox pair showed the highest reduction temperature 1392 °C; the reaction enthalpy is relatively high (560 kJ/kg) [36]. After three redox cycles, strong sintering of the powders occurred resulting in increasing reaction times. On the other hand, the high enthalpy, low cost material and its harmless behaviour makes iron oxide redox pair very interesting for applications where high reaction temperatures can be accepted. It should be also considered that redox equilibria temperatures could be lowered, when lower oxygen partial pressure than 0.21 atm can be used (Fig. 4).

The redox performance PbO_2/PbO was investigated by Müller et al. [39]. The author addressed a complex reaction reduction path, however, under none of the conditions applied could a reversible cycle be established also to all intermediates.

The research activities were extended towards the investigation of thermochemical behaviour of lead metal mixed oxides, a possible path for tuning relevant parameters such as reduction/oxidation temperature and cyclability [36, 38, 48–54].

Carrillo et [38] prepared manganese–cobalt pure and mixed oxides by precipitation with ammonia and investigated their redox cycle stability. As main conclusion, it was reported that Mn_2O_3 doped with cobalt result in thermal stabilization of the manganese oxide and a progressive loss of cyclability as the cobalt content increases and thus are not suitable to be a TCS material. Doping Co_2O_3 with manganese, no detrimental effect on the cyclability was observed. However, a slightly deteriorated kinetics, reduced heat stored/released amount and increased reaction temperatures render these doped oxides a less attractive alternative. The authors concluded that operation with pure oxides is more effective for TCS applications.

Block et al. [36] proposed cobalt-iron oxide and copper-cobalt oxide as promising binary materials for thermochemical storage thanks to their relatively high reaction enthalpies and reaction reduction rate with respect to the pure metal oxides. They found that manganese oxides containing iron or copper oxides are much less attractive due to the higher reduction temperature (999.2 °C) and almost unchanged reduction enthalpy (233 kJ/kg) with respect to pure oxide for the former and a decreased reduction enthalpy (157 kJ/kg) respect the pure Mn_2O_3 for the latter.

Recent research activity is exploiting, with very stimulating results, the thermochemical cycle based on solid oxide redox pairs through the realization of compact reactor/heat exchangers which conjugate heat adsorbed/released by the redox reaction and sensible heat [55–59]. Cascaded thermochemical storage studies represent also an interesting proof of concept that would allow to extend the temperature operation window [60].

5 Thermal Decomposition of Perovskites

Recently, perovskite oxides raised a growing interest as TCMs. The general formula for perovskite oxides is ABO_3, where A is generally a rare earth or an alkaline earth cation, while B is generally a transition metal cation. Their structure is ideally cubic with the A cations surrounded by 12 coordinated O anions, and B cations in the octahedral holes generated by the oxygen lattice. Perovskites undergo reversible non-stoichiometric reduction at specific temperature and oxygen partial pressure (pO_2), according to the following reaction:

$$ABO_{3(s)} + \text{Heat} \leftrightarrow ABO_{x-\delta(s)} + \frac{\delta}{2} O_{2(g)} \tag{6}$$

The oxygen release/uptake phenomena occurring in this material induce high heat effects, which mean that reduction is highly endothermic, while oxidation is highly exothermic, thus making perovskites interesting for TES applications. What makes these materials distinguishable with respect to other TES materials is their continuous, quasi-linear oxygen release/uptake (δ) within a very wide temperature range, thus potentially allowing the storage of heat at different temperature levels. Moreover, their reaction extent ($\Delta\delta$) depends on the particular conditions of temperature and oxygen partial pressure (see Fig. 6).

Encouraging but few preliminary results about perovskites as TES materials are reported in literature. In Table 1 some of the investigated perovskites compositions and their evaluated reaction enthalpy and extent are summarized.

Babiniek et al. [61] investigated $La_xSr_{1-x}Co_yMn_{1-y}O_3$ (LSCM) and $La_xSr_{1-x}Co_yFe_{1-y}O_3$ (LSCF) perovskites with $0.1 \le x \le 0.9$ and $0.1 \le y \le 0.9$. According to their studies, it seems that $\Delta\delta$ and the reduction onset temperature are

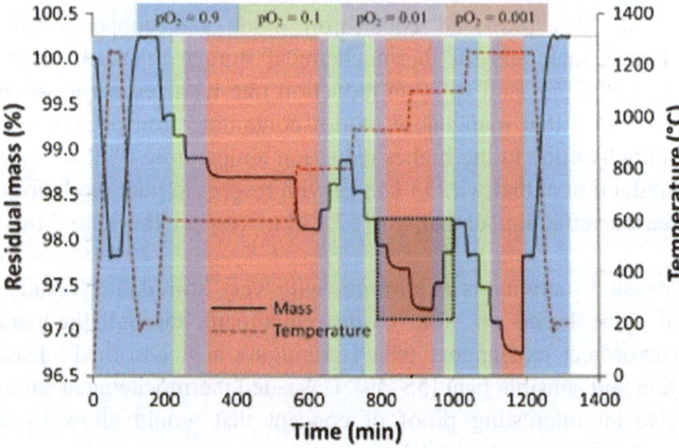

Fig. 6 From Ref. [61]. Reprinted with the permission of Elsevier

Table 1 Results on perovskites for TES applications reported in literature

Perovskite composition	Reaction enthalpy kJ/kg$_{ABO_3}$	δ	References
$La_{0.3}Sr_{0.7}Co_{0.8}Mn_{0.2}O_3$	250	0.4	[61]
$La_{0.2}Sr_{0.8}Co_{0.8}Fe_{0.2}O_3$	210	0.5	[61]
$La_{0.3}Sr_{0.7}Co_{0.9}Mn_{0.1}O_3$	245	0.45	[61]
$SrMnO_3$	17.4	0.06	[62]
$SrCoO_3$	39.6	0.33	[62]
$SrFeO_3$	53.3	0.24	[62]
$BaCoO_3$	292.1	0.47	[62]
$SrCo_{0.8}Fe_{0.2}O_3$	45.4	0.37	[62]
$SrCo_{0.2}Fe_{0.8}O_3$	24.0	0.36	[62]
$Ba_{0.5}Sr_{0.5}CoO_3$	172.6	0.45	[62]
$Ba_{0.5}Sr_{0.5}FeO_3$	21.4	0.27	[62]
$Ba_{0.5}Sr_{0.5}Co_{0.8}Fe_{0.2}O_3$	46.5	0.23	[62]
$Ba_{0.5}Sr_{0.5}Co_{0.2}Fe_{0.8}O_3$	21.2	0.26	[62]
$CaTi_{0.2}Mn_{0.8}O_3$	390	0.15	[63]
$CaAl_{0.2}Mn_{0.8}O_3$	370	0.2	[63]
$CaCr_{0.05}Mn_{0.95}O_3$	392	0.226	[64]
$Ca_{0.95}Sr_{0.05}MnO_3$	554	0.3	[64]
$Ca_{0.9}Sr_{0.1}MnO_3$	455	0.283	[64]

related to the crystallographic structure. Indeed, both perovskite families exhibited higher reduction extent and lower reduction onset for compositions with low La concentration, which adopt a cubic structure [61]. Among the investigated LSCM samples, the highest performances have been obtained for $La_{0.3}Sr_{0.7}Co_{0.8}Mn_{0.2}O_3$, $La_{0.3}Sr_{0.7}Co_{0.9}Mn_{0.1}O_3$ and $La_{0.2}Sr_{0.8}Co_{0.8}Fe_{0.2}O_3$ (reported in Table 1) with a reaction enthalpy equal to 250, 245 and 210 kJ/kg$_{ABO_3}$, respectively.

Zhang et al. [62] evaluated the potential of perovskites with Ba and/or Sr in A-site through the quantification of the oxidation reaction enthalpy by DSC analysis. The investigated compositions are listed in Table 1. Most of these compositions showed a reaction enthalpy well below 54 kJ/kg$_{ABO_3}$, except for $BaCoO_3$ and $Ba_{0.5}Sr_{0.5}CoO_3$, which exhibited a reaction enthalpy equal to 292 and 172 kJ/kg$_{ABO_3}$, respectively.

The above mentioned perovskite compositions utilize expensive and atomically heavy rare earth and/or metal cations that increase costs and decrease the mass-specific energy storage densities. For this reason, the attention has been focused on more cost-effective and least-massive perovskites compositions. $CaMnO_3$ perovskite for thermal heat storage application have been B-doped with cations that are difficult to reduce (such as Ti and Al) aiming at increasing the average M–O bond strength and thus the reduction enthalpy [63]. These dopants are also hypothesized to stabilize the structure, as $CaMnO_{3-\delta}$ is known to decompose in oxygen partial pressures <0.01 atm and at temperature around 1000 °C. $CaTi_{0.2}Mn_{0.8}O_{3-\delta}$ (CTM28) and $CaAl_{0.2}Mn_{0.8}O_{3-\delta}$ (CAM28) showed the most promising potential [63]. Indeed, from the equilibrium TGA data, the reaction enthalpy for CTM28 and CAM28, has been estimated to be 390 and 370 kJ/kg$_{ABO_3}$, respectively [63]. While, the reduction onset temperature in air for CTM28 and CAM28 is 900 and 750 °C, respectively [63].

The thermochemical behaviour of $CaCr_yMn_{1-y}O_3$ (y = 0.05 and 0.1) and of $Ca_{1-x}Sr_xMnO_3$ (x = 0.05 and 0.1) perovskites after undergoing heating from 500 to 1000 °C at $pO_2 = 10^{-4}$ atm has been evaluated by Imponenti et al. [64]. The A-site doped perovskite resulted in a higher reaction extent and higher reaction enthalpy. In particular, $Ca_{0.95}Sr_{0.05}MnO_3$ exhibits higher reaction enthalpy, equal to 554 kJ/kg$_{ABO_3}$.

It has to be underlined that the enthalpies values and the reaction extents here reported have been evaluated not only under different temperature and oxygen partial pressure conditions, but also with different methods (i.e. TGA, DSC) and numerical calculations. Indeed, from the investigation of $CaMn_{1-y}B_yO_{3-\delta}$ (B = Mg, Ti, Al) perovskites developed by Agrafiotis et al. [65], it is evident that lower reaction enthalpies for both $CaMnO_3$ and $CaMn_{0.9}Ti_{0.1}O_3$ (19/12 and 64/8 kJ/kg, respectively) are reported with respect to data from Babiniec et al. [63], likely due to different experimental conditions as the authors themselves pointed out.

In conclusion, only few studies specifically on perovskites as TES materials are available. What emerges from literature is the necessity to determine "standard" TGA experimental conditions (reduction/oxidation temperature, gas composition, heating/cooling rate) for comparing the different perovskite compositions, which is not easily obtainable since for each perovskite composition the experimental conditions (pO_2, T) should be optimized aiming at obtaining the highest reaction performances (in terms of reaction enthalpy). As a main issue, cost-effective and least-massive perovskite composition with high reaction enthalpy (at least comparable with that of the most performing binary oxide systems) in a wide operational temperature range has still to be determined. However, the extensive literature about the oxygen non-stoichiometry of perovskites and their high tunability give a considerable scope for further investigation of this class of oxides.

References

1. Shkatulov A, Aristov Y (2015) Modification of magnesium and calcium hydroxides with salts: an efficient way to advanced materials for storage of middle-temperature heat. Energy 85:667–676. https://doi.org/10.1016/J.ENERGY.2015.04.004
2. Scapino L, Zondag HA, Van Bael J et al (2017) Sorption heat storage for long-term low-temperature applications: a review on the advancements at material and prototype scale. Appl Energy 190:920–948. https://doi.org/10.1016/j.apenergy.2016.12.148
3. Yan T, Wang RZ, Li TX et al (2015) A review of promising candidate reactions for chemical heat storage. Renew Sustain Energy Rev 43:13–31. https://doi.org/10.1016/J.RSER.2014.11.015

4. Trausel F, de Jong A-J, Cuypers R (2014) A review on the properties of salt hydrates for thermochemical storage. Energy Procedia 48:447–452. https://doi.org/10.1016/J.EGYPRO.2014.02.053
5. Huang Q, Lu G, Wang J, Yu J (2011) Thermal decomposition mechanisms of $MgCl_2{\cdot}6H_2O$ and $MgCl_2{\cdot}H_2O$. J Anal Appl Pyrolysis 91:159–164. https://doi.org/10.1016/J.JAAP.2011.02.005
6. N'Tsoukpoe KE, Rammelberg HU, Lele AF et al (2015) A review on the use of calcium chloride in applied thermal engineering. Appl Therm Eng 75:513–531. https://doi.org/10.1016/J.APPLTHERMALENG.2014.09.047
7. van Essen VM, Zondag HA, Gores JC et al (2009) Characterization of $MgSO_4$ hydrate for thermochemical seasonal heat storage. J Sol Energy Eng 131:41014. https://doi.org/10.1115/1.4000275
8. Kallenberger PA, Brieler FJ, Posern K, Fröba M (2016) Magnesium sulfate/polymer composites for seasonal, thermochemical energy storage. Chemie Ing Tech 88:379–384. https://doi.org/10.1002/cite.201500095
9. Posern K, Linnow K, Niermann M et al (2015) Thermochemical investigation of the water uptake behavior of $MgSO_4$ hydrates in host materials with different pore size. Thermochim Acta 611:1–9. https://doi.org/10.1016/J.TCA.2015.04.031
10. Korhammer K, Druske M-M, Fopah-Lele A et al (2016) Sorption and thermal characterization of composite materials based on chlorides for thermal energy storage. Appl Energy 162:1462–1472. https://doi.org/10.1016/J.APENERGY.2015.08.037
11. Gaeini M, Rouws AL, Salari JWO et al (2018) Characterization of microencapsulated and impregnated porous host materials based on calcium chloride for thermochemical energy storage. Appl Energy 212:1165–1177. https://doi.org/10.1016/j.apenergy.2017.12.131
12. Gutierrez A, Ushak S, Mamani V et al (2017) Characterization of wastes based on inorganic double salt hydrates as potential thermal energy storage materials. Sol Energy Mater Sol Cells 170:149–159. https://doi.org/10.1016/J.SOLMAT.2017.05.036
13. Rammelberg HU, Osterland T, Priehs B et al (2016) Thermochemical heat storage materials—performance of mixed salt hydrates. Sol Energy 136:571–589. https://doi.org/10.1016/J.SOLENER.2016.07.016
14. Knoll C, Müller D, Artner W et al (2017) Probing cycle stability and reversibility in thermochemical energy storage—$CaC_2O_4{\cdot}H_2O$ as perfect match? Appl Energy 187:1–9. https://doi.org/10.1016/J.APENERGY.2016.11.053
15. N'Tsoukpoe KE, Schmidt T, Rammelberg HU et al (2014) A systematic multi-step screening of numerous salt hydrates for low temperature thermochemical energy storage. Appl Energy 124:1–16. https://doi.org/10.1016/J.APENERGY.2014.02.053
16. Deutsch M, Müller D, Aumeyr C et al (2016) Systematic search algorithm for potential thermochemical energy storage systems. Appl Energy 183:113–120. https://doi.org/10.1016/J.APENERGY.2016.08.142
17. Donkers PAJ, Sögütoglu LC, Huinink HP et al (2017) A review of salt hydrates for seasonal heat storage in domestic applications. Appl Energy 199:45–68. https://doi.org/10.1016/J.APENERGY.2017.04.080
18. Kato Y, Koyama M, Fukushima Y, Nakagaki T (2016) Energy technology roadmaps of Japan. Springer, Tokyo
19. Schaube F, Koch L, Wörner A, Müller-Steinhagen H (2012) A thermodynamic and kinetic study of the de- and rehydration of $Ca(OH)_2$ at high H_2O partial pressures for thermo-chemical heat storage. Thermochim Acta 538:9–20. https://doi.org/10.1016/J.TCA.2012.03.003
20. Ogura H, Kubota M, Suzuki H, Yamakawa T (2009) Fundamental experimental study on chemical heat pump for storing low-temperature waste heat and releasing cold-heat. Kagaku Kogaku Ronbunshu 35:506–510. https://doi.org/10.1252/kakoronbunshu.35.506
21. Hirata Y, Fujioka K, Fujiki S (2003) Preparation of fine particles of calcium chloride with expanded graphite for enhancement of the driving reaction for chemical heat pumps. J Chem Eng Japan 36:827–832. https://doi.org/10.1252/jcej.36.827

22. Esaki T, Kobayashi N (2016) Reaction rate characteristics of $SrBr_2$ hydration system for chemical heat pump cooling mode. J Mater Sci Chem Eng 4:106–115. https://doi.org/10.4236/msce.2016.42012
23. Piperopoulos E, Mastronardo E, Fazio M et al (2018) Enhancing the volumetric heat storage capacity of Mg(OH)2 by the addition of a cationic surfactant during its synthesis. Appl Energy 215:512–522. https://doi.org/10.1016/J.APENERGY.2018.02.047
24. Kato Y, Yamashita N, Kobayashi K, Yoshizawa Y (1996) Kinetic study of the hydration of magnesium oxide for a chemical heat pump. Appl Therm Eng 16:853–862. https://doi.org/10.1016/1359-4311(96)00009-9
25. Ishitobi H, Hirao N, Ryu J, Kato Y (2013) Evaluation of heat output densities of lithium chloride-modified magnesium hydroxide for thermochemical energy storage. Ind Eng Chem Res 52:5321–5325. https://doi.org/10.1021/ie302841y
26. Myagmarjav O, Ryu J, Kato Y (2014) Lithium bromide-mediated reaction performance enhancement of a chemical heat-storage material for magnesium oxide/water chemical heat pumps. Appl Therm Eng 63:170–176. https://doi.org/10.1016/J.APPLTHERMALENG.2013.10.045
27. Zamengo M, Ryu J, Kato Y (2013) Magnesium hydroxide—expanded graphite composite pellets for a packed bed reactor chemical heat pump. Appl Therm Eng 61:853–858. https://doi.org/10.1016/J.APPLTHERMALENG.2013.04.045
28. Mastronardo E, Bonaccorsi L, Kato Y et al (2017) Strategies for the enhancement of heat storage materials performances for $MgO/H_2O/Mg(OH)_2$ thermochemical storage system. Appl Therm Eng 120:626–634. https://doi.org/10.1016/J.APPLTHERMALENG.2017.04.004
29. Mastronardo E, Bonaccorsi L, Kato Y et al (2016) Efficiency improvement of heat storage materials for $MgO/H_2O/Mg(OH)_2$ chemical heat pumps. Appl Energy 162:31–39. https://doi.org/10.1016/J.APENERGY.2015.10.066
30. Tae Kim S, Kato Y (2011) Reactivity enhancement of chemical materials used in packed bed reactor of chemical heat pump. Prog Nucl Energy 53:1027–1033. https://doi.org/10.1016/J.PNUCENE.2011.05.013
31. Zamengo M, Tomaškovic J, Ryu J, Kato Y (2016) Thermal conductivity measurements of expanded graphite-magnesium hydroxide composites for packed bed reactors of chemical heat storage/pump systems. J Chem Eng Japan 49:261–267. https://doi.org/10.1252/jcej.14we292
32. Mastronardo E, Kato Y, Bonaccorsi L et al (2017) Thermochemical storage of middle temperature wasted heat by functionalized $C/Mg(OH)_2$ hybrid materials. Energies 10:70. https://doi.org/10.3390/en10010070
33. Mastronardo E, Bonaccorsi L, Kato Y et al (2016) Thermochemical performance of carbon nanotubes based hybrid materials for $MgO/H2O/Mg(OH)_2$ chemical heat pumps. Appl Energy 181:232–243. https://doi.org/10.1016/J.APENERGY.2016.08.041
34. Prieto C, Cooper P, Fernández AI, Cabeza LF (2016) Review of technology: thermochemical energy storage for concentrated solar power plants. Renew Sustain Energy Rev 60:909–929. https://doi.org/10.1016/J.RSER.2015.12.364
35. Bulfin B, Vieten J, Agrafiotis C et al (2017) Applications and limitations of two step metal oxide thermochemical redox cycles; a review. J Mater Chem A 5:18951–18966. https://doi.org/10.1039/C7TA05025A
36. Block T, Schmücker M (2016) Metal oxides for thermochemical energy storage: a comparison of several metal oxide systems. Sol Energy 126:195–207. https://doi.org/10.1016/J.SOLENER.2015.12.032
37. Agrafiotis C, Roeb M, Schmücker M, Sattler C (2014) Exploitation of thermochemical cycles based on solid oxide redox systems for thermochemical storage of solar heat. Part 1: testing of cobalt oxide-based powders. Sol Energy 102:189–211. https://doi.org/10.1016/J.SOLENER.2013.12.032
38. Carrillo AJ, Moya J, Bayón A et al (2014) Thermochemical energy storage at high temperature via redox cycles of Mn and Co oxides: pure oxides versus mixed ones. Sol Energy Mater Sol Cells 123:47–57. https://doi.org/10.1016/J.SOLMAT.2013.12.018

39. Müller D, Knoll C, Artner W et al (2017) Combining in-situ X-ray diffraction with thermogravimetry and differential scanning calorimetry—an investigation of Co_3O_4, MnO_2 and PbO_2 for thermochemical energy storage. Sol Energy 153:11–24. https://doi.org/10.1016/J.SOLENER.2017.05.037
40. Schrader AJ, Muroyama AP, Loutzenhiser PG (2015) Solar electricity via an Air Brayton cycle with an integrated two-step thermochemical cycle for heat storage based on Co3O4/CoO redox reactions: thermodynamic analysis. Sol Energy 118:485–495. https://doi.org/10.1016/J.SOLENER.2015.05.045
41. Muroyama AP, Schrader AJ, Loutzenhiser PG (2015) Solar electricity via an Air Brayton cycle with an integrated two-step thermochemical cycle for heat storage based on Co_3O_4/CoO redox reactions II: kinetic analyses. Sol Energy 122:409–418. https://doi.org/10.1016/J.SOLENER.2015.08.038
42. Schrader AJ, De Dominicis G, Schieber GL, Loutzenhiser PG (2017) Solar electricity via an Air Brayton cycle with an integrated two-step thermochemical cycle for heat storage based on Co_3O_4/CoO redox reactions III: solar thermochemical reactor design and modeling. Sol Energy 150:584–595. https://doi.org/10.1016/J.SOLENER.2017.05.003
43. Singh A, Tescari S, Lantin G et al (2017) Solar thermochemical heat storage via the Co_3O_4/CoO looping cycle: storage reactor modelling and experimental validation. Sol Energy 144:453–465. https://doi.org/10.1016/J.SOLENER.2017.01.052
44. Carrillo AJ, Serrano DP, Pizarro P, Coronado JM (2014) Thermochemical heat storage based on the Mn_2O_3/Mn_3O_4 redox couple: influence of the initial particle size on the morphological evolution and cyclability. J Mater Chem A 2:19435–19443. https://doi.org/10.1039/C4TA03409K
45. Deutsch M, Horvath F, Knoll C et al (2017) High-temperature energy storage: kinetic investigations of the CuO/Cu_2O reaction cycle. Energy Fuels 31:2324–2334. https://doi.org/10.1021/acs.energyfuels.6b02343
46. Haseli P, Jafarian M, Nathan GJ (2017) High temperature solar thermochemical process for production of stored energy and oxygen based on CuO/Cu_2O redox reactions. Sol Energy 153:1–10. https://doi.org/10.1016/J.SOLENER.2017.05.025
47. Alonso E, Pérez-Rábago C, Licurgo J et al (2015) First experimental studies of solar redox reactions of copper oxides for thermochemical energy storage. Sol Energy 115:297–305. https://doi.org/10.1016/J.SOLENER.2015.03.005
48. Wokon M, Block T, Nicolai S et al (2017) Thermodynamic and kinetic investigation of a technical grade manganese-iron binary oxide for thermochemical energy storage. Sol Energy 153:471–485. https://doi.org/10.1016/J.SOLENER.2017.05.045
49. André L, Abanades S, Cassayre L (2017) High-temperature thermochemical energy storage based on redox reactions using Co-Fe and Mn-Fe mixed metal oxides. J Solid State Chem 253:6–14. https://doi.org/10.1016/J.JSSC.2017.05.015
50. Wokon M, Kohzer A, Linder M (2017) Investigations on thermochemical energy storage based on technical grade manganese-iron oxide in a lab-scale packed bed reactor. Sol Energy 153:200–214. https://doi.org/10.1016/J.SOLENER.2017.05.034
51. Carrillo AJ, Serrano DP, Pizarro P, Coronado JM (2016) Manganese oxide-based thermochemical energy storage: modulating temperatures of redox cycles by Fe–Cu co-doping. J Energy Storage 5:169–176. https://doi.org/10.1016/J.EST.2015.12.005
52. Carrillo AJ, Serrano DP, Pizarro P, Coronado JM (2016) Understanding redox kinetics of iron-doped manganese oxides for high temperature thermochemical energy storage. J Phys Chem C 120:27800–27812. https://doi.org/10.1021/acs.jpcc.6b08708
53. Carrillo AJ, Serrano DP, Pizarro P, Coronado JM (2015) Improving the thermochemical energy storage performance of the Mn_2O_3 /Mn_3O_4 redox couple by the incorporation of iron. Chemsuschem 8:1947–1954. https://doi.org/10.1002/cssc.201500148
54. Block T, Knoblauch N, Schmücker M (2014) The cobalt-oxide/iron-oxide binary system for use as high temperature thermochemical energy storage material. Thermochim Acta 577:25–32. https://doi.org/10.1016/J.TCA.2013.11.025

55. Tescari S, Singh A, Agrafiotis C et al (2017) Experimental evaluation of a pilot-scale thermochemical storage system for a concentrated solar power plant. Appl Energy 189:66–75. https://doi.org/10.1016/J.APENERGY.2016.12.032
56. Agrafiotis C, Becker A, Roeb M, Sattler C (2016) Exploitation of thermochemical cycles based on solid oxide redox systems for thermochemical storage of solar heat. Part 5: testing of porous ceramic honeycomb and foam cascades based on cobalt and manganese oxides for hybrid sensible/thermochemical heat storage. Sol Energy 139:676–694. https://doi.org/10.1016/J.SOLENER.2016.09.013
57. Karagiannakis G, Pagkoura C, Halevas E et al (2016) Cobalt/cobaltous oxide based honeycombs for thermochemical heat storage in future concentrated solar power installations: multi-cyclic assessment and semi-quantitative heat effects estimations. Sol Energy 133:394–407. https://doi.org/10.1016/J.SOLENER.2016.04.032
58. Agrafiotis C, Tescari S, Roeb M et al (2015) Exploitation of thermochemical cycles based on solid oxide redox systems for thermochemical storage of solar heat. Part 3: cobalt oxide monolithic porous structures as integrated thermochemical reactors/heat exchangers. Sol Energy 114:459–475. https://doi.org/10.1016/J.SOLENER.2014.12.037
59. Agrafiotis C, Roeb M, Schmücker M, Sattler C (2015) Exploitation of thermochemical cycles based on solid oxide redox systems for thermochemical storage of solar heat. Part 2: redox oxide-coated porous ceramic structures as integrated thermochemical reactors/heat exchangers. Sol Energy 114:440–458. https://doi.org/10.1016/J.SOLENER.2014.12.036
60. Agrafiotis C, Roeb M, Sattler C (2016) Exploitation of thermochemical cycles based on solid oxide redox systems for thermochemical storage of solar heat. Part 4: screening of oxides for use in cascaded thermochemical storage concepts. Sol Energy 139:695–710. https://doi.org/10.1016/J.SOLENER.2016.04.034
61. Babiniec SM, Coker EN, Miller JE, Ambrosini A (2015) Investigation of $La_xSr_{1-x}Co_yM_{1-y}O_{3-\delta}$ (M = Mn, Fe) perovskite materials as thermochemical energy storage media. Sol Energy 118:451–459. https://doi.org/10.1016/j.solener.2015.05.040
62. Zhang Z, Andre L, Abanades S (2016) Experimental assessment of oxygen exchange capacity and thermochemical redox cycle behavior of Ba and Sr series perovskites for solar energy storage. Sol Energy 134:494–502. https://doi.org/10.1016/j.solener.2016.05.031
63. Babiniec SM, Coker EN, Miller JE, Ambrosini A (2015) Doped calcium manganites for advanced high-temperature thermochemical energy storage. 6–10. https://doi.org/10.1002/er.3467
64. Imponenti L, Albrecht KJ, Braun RJ, Jackson GS (2016) Measuring thermochemical energy storage capacity with redox cycles of doped-$CaMnO_3$. ECS Trans 72:11–22. https://doi.org/10.1149/07207.0011ecst
65. Agrafiotis C, Block T, Senholdt M et al (2017) Exploitation of thermochemical cycles based on solid oxide redox systems for thermochemical storage of solar heat. Part 6: testing of Mn-based combined oxides and porous structures. Sol Energy 149:227–244. https://doi.org/10.1016/j.solener.2017.03.083

Sorption Thermal Energy Storage

Andrea Frazzica, Vincenza Brancato, Valeria Palomba and Salvatore Vasta

Abstract In the present chapter, an introduction about the concept of sorption TES technology is reported. The closed and open configurations are discussed and an overview on the ongoing research and development activities for materials, components and systems is given.

1 Introduction

Sorption TES belongs to the wider class of thermochemical energy storage. The definition 'sorption' was first reported by McBain [1] in 1909 to describe the interaction occurring between a gaseous sorbate and a liquid or solid sorbent. The former interaction was identified as absorption while the latter as adsorption. Absorption was defined as '*a modified solid solution in which practically only the outer layers become saturated, owing to the difficulty of diffusion in solids*', while adsorption was defined as '*condensation on the outside of the surface of the solid*'. In most of the TESs based on sorption technology, the sorbate is represented by water vapour. For this reason, a classification among the materials for sorption TES is mainly referred to the sorbent employed, as represented in Fig. 1. The classification is performed taking into account the main reaction occurring between the sorbate and the sorbent materials. The absorption into liquids relates to absorption of water vapour inside salt solutions, or ammonia inside liquid water. The solid adsorption is related to the surface interaction between the sorbate (i.e. water vapour) with the solid surface, through weak bonds, like Van der Waals and hydrogen bonds. Nevertheless, in some cases different phenomena can coexist. In particular, the sorption composites present an intermediate behaviour between physical adsorption onto the solid (i.e. the host matrix) and chemical hydration reaction (i.e. in the embedded salt).

A. Frazzica (✉) · V. Brancato · V. Palomba · S. Vasta
CNR-ITAE, Istituto di Tecnologie Avanzate per l'Energia "Nicola Giordano",
Via Salita S. Lucia sopra Contesse 5, 98126 Santa Lucia, Messina, Italy
e-mail: andrea.frazzica@itae.cnr.it

A. Frazzica and L. F. Cabeza (eds.), *Recent Advancements in Materials and Systems for Thermal Energy Storage*, Green Energy and Technology,
https://doi.org/10.1007/978-3-319-96640-3_4

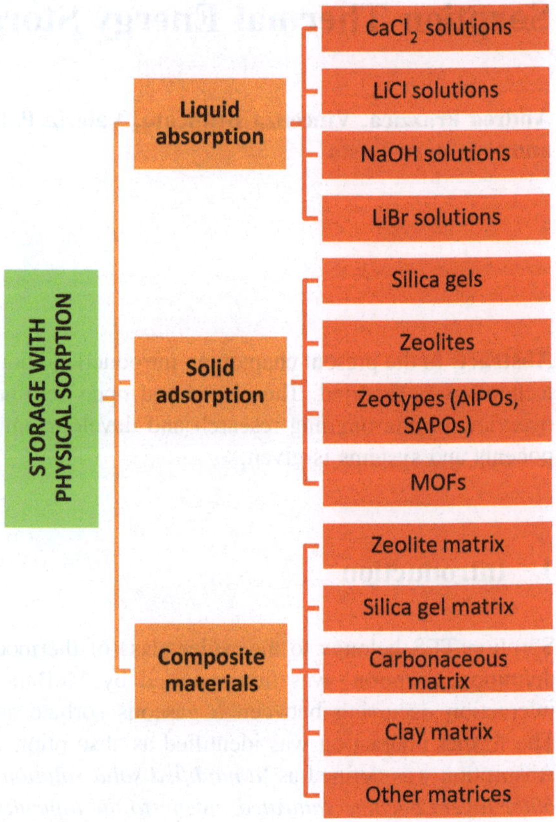

Fig. 1 Categorisation of sorbent materials for sorption TES

Generally, there are two system configurations for sorption TES: closed and open cycle. Figure 2 reports the working phases of a closed sorption TES. During charging phase, the reactor, in which the sorbent material is saturated of sorbate (e.g. water), is regenerated exploiting heat coming from the heat source, Q_{des}. The desorbed vapour is then condensed in the condenser, and the heat of condensation, Q_{cond}, is either dissipated in the ambient or delivered to the load, if the temperature level is sufficient to satisfy it. Once the charging process is completed (i.e. the sorbent material is almost dry), the connection between the condenser and the reactor is closed. Under this condition, the system can keep the stored energy for indefinite time, since the thermal energy is stored as sorption potential between sorbate and sorbent material. In order to get back the stored thermal energy, the connection between liquid sorbate reservoir, which in this phase acts as evaporator and sorbent material in the reactor is again opened. During this discharging phase, the sorbate is evaporated by means of the heat from the ambient, Q_{evap}, then the

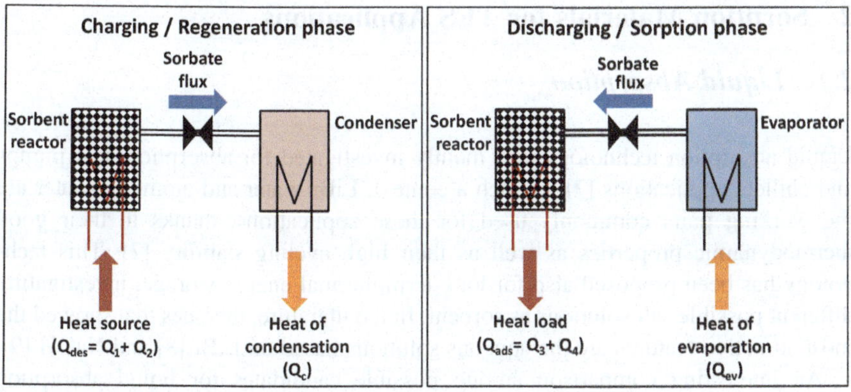

Fig. 2 Closed sorption TES cycle: charging and discharging phase

vapour fluxes to the sorbent material, since the sorption process is exothermic, heat is released to the load, Q_{sorp}. Clearly, this process is defined closed since the sorbate is continuously condensed/evaporated in a closed system without any mass exchange with the ambient.

Differently, the open sorption TES system, represented in Fig. 3, continuously exchanges mass (sorbate) with the ambient. Actually, the two charging/discharging phases are similar to the ones already described for the closed cycle. Main difference is that, in this case, heat is provided and extracted by fluxing air through the sorbent material contained in the reactor. Particularly, during charging/regeneration phase, a hot and dry air flux enters the storage, causing the desorption of sorbate (i.e. water), and exiting at lower temperature and higher humidity content. During discharging/sorption phase, the humid and cooled air flux is provided to the dry sorbent material, which triggers the sorption and the consequent release of stored heat, which reflects on the exiting hot and dry air flux.

In the literature, several reviews focus on the analysis of the peculiarities of each class of sorbent materials for TES applications, some suggested readings are [2–6]. The following sections focus on the most recent advancements achieved in the sorption TES technology during last years.

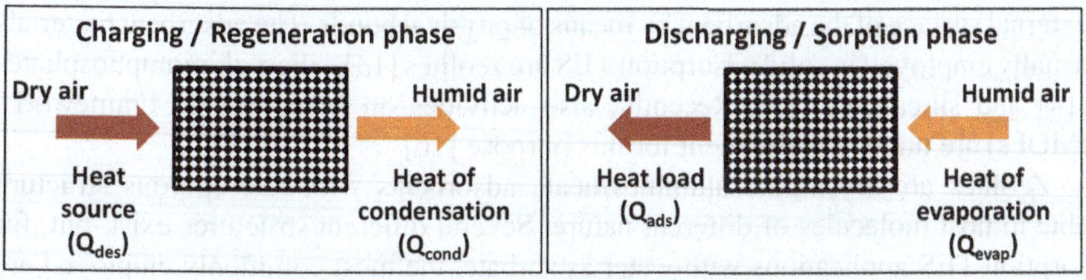

Fig. 3 Open sorption TES cycle: charging and discharging phase

2 Sorption Materials for TES Applications

2.1 Liquid Absorption

Liquid absorption technology was mainly investigated for absorption heat pumps and chillers applications [7]. In such a context, LiBr-water and ammonia-water are the working pairs commonly used for these applications, thanks to their good thermodynamic properties as well as their high cycling stability [7]. This technology has been proposed also for long-term thermal energy storage, investigating different possible salt solutions as sorbent. In the literature, the ones that showed the most attractive features are the aqueous solutions based on LiBr [8] and NaOH [9].

An interesting comparison among possible candidates for liquid absorption thermal energy storage was presented by [10]. They investigated seven working pairs, (aqueous solutions of $CaCl_2$, Glycerine, KOH, LiBr, LiCl, NaOH and water/ammonia working pairs) analysing also the effect of allowing partial crystallisation of the salt inside the solution. In the paper, different features were analysed, taking into account optimal charging temperatures, cost of the materials and corrosion resistance. In Table 1 the achievable TES densities, minimum charging temperatures required, the storage system efficiency and the absolute pressure of the system are summarised. The data are reported for both solutions working without any crystallisation and solutions achieving a crystallisation up to a solid salt mass four times higher than the salt solution mass.

As can be highlighted in Table 1, when no crystallisation is allowed, the most attractive solution is the one employing LiCl as salt, presenting also low charging temperature and good efficiency. Similar result is obtained when the crystallisation happens. Clearly, neither glycerine/water nor water/ammonia working pairs can allow any crystallisation under the operating conditions typical of the TES. Generally, crystallisation improves the storage density, lowering at the same time the efficiency of the system. More discussions on the proper selection of liquid absorption working pairs can be found in [11, 12].

2.2 Solid Adsorption

In the solid adsorption TES, the adsorbate molecules interact with the solid on the external surface of the adsorbent by means of physical bonds. The adsorbent materials usually employed in solid adsorption TES are zeolites [13], silico-aluminophosphates [14] and silica gels [15]. Recently, also activities on Metalorganic Frameworks (MOFs) are under development for this purpose [16].

Zeolites are crystalline aluminosilicate adsorbents with microporous structure able to host molecules of different nature. Several different structures exist, but, for sorption TES applications with water as sorbate, the most commonly employed are 4A, 54, 13X and Y [17]. These materials are quite well established and investigated

Table 1 Main features of investigated working pairs for liquid absorption TES

Crystallisation ratio	Parameter	$CaCl_2/H_2O$	Glycerine/H_2O	KOH/H_2O	$LiBr/H_2O$	$LiCl/H_2O$	$NaOH/H_2O$	H_2O/NH_3
0	Storage density (kWh/kg)	0.254	0.053	0.727	0.56	1.218	0.432	0.365
	Minimum charging temperature (°C)	44.8	53.0	63.0	72.0	65.6	50.0	155.5
	Efficiency	0.909	0.545	0.830	0.850	0.950	0.750	0.665
	Absolute pressure (kPa)	1.2–4.2	1.2–4.2	1.2–4.2	1.2–4.2	1.2–4.2	1.2–4.2	615–1167
4	Storage density (kWh/kg)	0.562	–	0.916	0.723	1.425	0.661	–
	Minimum charging temperature (°C)	54.2	–	84.0	93.3	77.6	57.0	–
	Efficiency	0.885	–	0.586	0.835	0.923	0.767	–
	Absolute pressure (kPa)	1.2–4.2	–	1.2–4.2	1.2–4.2	1.2–4.2	1.2–4.2	–

Adapted from [10]

in the literature. Nevertheless, still some activities are currently performed. For instance, in [18], a comparison between commercial 13X with Na^+ as cation and exchanged samples with Mg^{2+} and Ca^{2+} was performed. It demonstrated that the effect of the cation allows obtaining higher storage density when the Mg^{2+} cations replace the Na^+ in the zeolite structure. Another analysis performed on the exchanged cations in 13X samples was recently reported in [19]. In this case, different cations were employed, demonstrating that the increasing trend of the storage density depends on the exchanged cations (i.e. Zn > Sr > Cd > Na). This confirms the possibility of tuning it by properly varying the exchanged cations. In general, zeolites are characterised by high charging temperatures (i.e. above 150 °C), which make them suitable for industrial TES applications but less attractive for low temperature solar TES.

Aluminophosphates (AlPOs) and other derived structures like silico-aluminophosphates (SAPOs) are characterised by a crystalline and porous structure, similar to the classical zeolites. For this reason they are often referred as zeotypes [20, 21]. These materials are usually characterised by an adsorption isotherm V according to IUPAC classification [22], showing a partially hydrophobic behaviour at low relative pressure and a partially hydrophilic behaviour at high relative pressure. This isotherm is sometimes defined as 'S-shaped' since a high water vapour exchange can be achieved in a narrow relative pressure range, making these materials suitable for sorption TES applications [21, 23]. Recently, Brancato and Frazzica [21] comparatively analysed the achievable performance of three among the most attractive zeotypes for adsorption heat transformation, namely, AlPO-18, SAPO-34 and FAPO-34. In particular, for TES application, they demonstrated that the storage density achievable by these materials, when heat source at 90 °C is available, can reach up to 0.14 kWh/kg, to provide space heating, which represents a promising performance indicator for sorption TES application. Nevertheless, the available micro pores volume typical of these materials, which cannot allow further increasing the sorption capacity and thus the related TES density, limits this value. Some other papers dealing with the analysis of these materials are the following [14, 24].

Silica gels are porous materials with an amorphous structure, which have been historically considered for sorption TES applications, thanks to their ability as water sorbents. These materials can be synthesised under different conditions, achieving porous structure from the micro- to meso-pores [20]. They attracted a lot of attention, especially thanks to their low cost and wide commercial availability. Nevertheless, the interest towards this class of materials has been recently reduced, since outcomes of some research projects, like Modestore [25], demonstrated that they have low water vapour exchange, which limits the achievable TES density.

2.3 Composite Materials

The development of composite sorbent materials was proposed to overcome some of the main issues of the pure salt applied for sorption TES (i.e. cycling stability, low thermal conductivity and vapour transfer resistance) [26, 27]. Indeed, in these materials the salt is embedded inside a porous matrix, which disperses the salt, allowing lowering mass transfer resistance and limiting agglomeration problems. The research activity on the development and testing of this class of materials is gaining a lot of attention, thanks to high achievable TES density as well as their flexibility in terms of composition that makes them adaptable for different operating conditions.

Several salts coupled to different porous matrixes (e.g. zeolites, silica gels, carbonaceous materials) were investigated and reported in the literature. A deep literature review is reported in [6].

In Table 2, an overview of the composite sorbents developed and tested so far is reported. As can be seen, also from the storage density point of view, these materials show an intermediate behaviour between pure salts and physical adsorbents.

Table 3 reports a comparison among the average thermo-physical parameters of different sorbent materials for TES applications. As can be seen, the composites present the most attractive performance in terms of sorption capacity, which means achievable TES density. While, among the pure adsorbents, MOFs present attractive features. Nevertheless, both categories presents some open issues to be investigated, such as the cycling stability and the cost, especially for MOFs. This confirms the needs for further research and development activities.

3 Components and Systems for Sorption TES Application

Material development represents only the first stage in the design of a TES system, since the realisation of a complete system generally involves several technical challenges. Research on sorption storages, useful for application in the residential sector, still requires significant efforts for increasing the technological level. Nonetheless, during the last years, significant efforts were devoted to the study of components and prototypes, demonstrating the practical feasibility of the technology. To present a complete overview of the most recent systems reported in the literature, Table 4 summarises the relevant features and findings regarding the prototypes, while Table 5 presents the testing conditions and main experimental outcomes of a vast amount of prototypes of thermochemical storages available in the literature.

Among absorption systems, a wide experimental activity was performed on NaOH/water working pair for the development of a pilot demonstrator [42]. However, the results of the experimental campaign, despite proving the successful

Table 2 Some of the most interesting composite sorbents reported in the literature for sorption TES applications

Matrixes	Salts	Salt (wt %)	Experimental conditions: temperature discharging/charging range (°C)	Experimental conditions: pressure range (kPa)	TES density ($kWh/m^3_{adsorbent}$) or ($kWh/kg_{adsorbent}$)	References
Silica gel, zeolite 13X, vermiculite	$CaCl_2$, $MgSO_4$, $Ca(NO_3)_2$, $LiNO_3$, LiBr	2–65	30–140	–	Up to 50 (kWh/m^3)	[28]
Expanded natural graphite	KCl, $CaCl_2$	63–90	25–200	2.0	Up to 175 (kWh/m^3)	[29]
Siliceous shale	LiCl, $CaCl_2$	9.6	15–150	2.21	55.5 (kWh/m^3)	[30]
Graphite, zeolite A, sand	$MgCl_2$	46–69	35–200	1.18–2.18	155.5 (kWh/m^3)	[31]
Activated carbon, silica solution, expanded graphite	LiCl	32–45	30–90	2.54	200–397.2 (kWh/m^3)	[32]
Silica gel	$CaCl_2$	43	50–100	0–5.0	–	[33]
Zeolite 13X, silica gel	$MgSO_4$	15	25–150	2.15	166.6 (kWh/m^3)	[34]
Silica gel, aluminosilicate	$MgCl_2$, $CaCl_2$	30	–	–	166.6 (kWh/m^3)	[35]
Attapulgite	$CaCl_2$	30	40–400	–	113.8–300 (kWh/m^3)	[36]
Attapulgite	$MgSO_4$, $MgCl_2$	33	30–130	3.6	0.24 (kWh/kg)	[37]
FeKIL2	$CaCl_2$	7	40–150	1.18–5.53	0.15 (kWh/kg)	[38]
Expanded natural graphite, activated carbon foam	KCl, $CaCl_2$	31–90	25–200	1–2	0.36–0.40 (kWh/kg)	[39]
Multi Wall Carbon Nanotubes	LiCl, $CaCl_2$	44–53	35–75	0.87–5.6	0.14–0.47 (kWh/kg)	[40]
Vermiculite	LiCl	59	35–85	0.87–5.6	0.5–0.72 (kWh/kg)	[41]

Adapted from [6]

Table 3 Thermo-physical parameters of different sorbent materials for TES applications

	Silica gels	Zeolites	AlPOs/SAPOs
Adsorption heat (kJ/kg)	160–180[a]	50–300[a]	250–300[a]
Typical desorption temperatures (°C)	50–80	70–350	60–90
Density (kg/m^3)	650–700	650–900	800–900
Specific heat (kJ/kg K)	0.8–0.9	0.85–0.95	0.85–0.95
Thermal conductivity (W/mK)	0.15–0.20	0.15–0.25	0.15–0.25
Possible working fluids	Water	Water	Water
Amount of uptake exchanged in a typical cycle (g/g)	0.03–0.10	up to 0.2	up to 0.25
	Composites	MOFs	Activated carbons
Adsorption heat (kJ\kg)	50–250[a]	20–200[b]	45–900[c]
Typical desorption temperatures (°C)	60–90	60–150	80–200
Density (kg/m^3)	300–600	1000–2000	700–750
Specific heat (kJ/kg K)	0.95–1.05	0.8–1.2	0.8–1.5
Thermal conductivity (W/mK)	0.15–0.30	0.10–015	0.15–0.75
Possible working fluids	Water, methanol, ethanol	Water, methanol, ethanol	Methanol, ethanol, ammonia
Amount of uptake exchanged in a typical cycle (g/g)	Up to 0.8	0.16–1.10	015–0.60

[a]the heat of adsorption is calculated for a cycle with T_{des} = 100 °C, T_{cond} = 30 °C, T_{ads} = 50 °C, T_{ev} = 10 °C, with water as sorbate
[b]the heat of adsorption is calculated from isotherms at 298, 303 and 333 K, with water as sorbate
[c]the range of heat of adsorption is calculated with methanol and ammonia as sorbates

operation of the system, evidenced that a proper design of the components is needed to avoid low thermal power output.

In the field of solid adsorption, several prototypes were developed for space heating application, employing zeolite 13X as adsorbent within open cycles. The main peculiarity of the prototypes presented in [43–45] is their modularity, which allows parallel or series operation, to extend the time during which the useful effect is produced or increase the thermal power output. Instead, only a few prototypes were developed using zeolite (i.e. FAM Z02) for closed adsorption applications in space heating and cooling, which have further stressed the need for an optimised design of the storage unit itself and the auxiliary components, such as the evaporator.

On the contrary, during the last years, the intense activity on composite materials lead to the development of an increasing number of prototypes. Apart from the tailoring of the composite for the specific application, the main peculiarity of several of these prototypes is the use under different conditions than the standard charge/discharge cycle presented in Fig. 2. For instance, the prototypes developed

Table 4 Sorption storage systems recently presented

References	System description	Classification	Working pair	Size	Relevant results
[42]	Closed sorption heat storage for seasonal storage applications	Liquid absorption	NaOH/water	Pilot demonstrator	About 1 kW useful effect output was produced. Results of tests highlighted the need for a careful design of HEXs and the possibility to use a different solution rather than tube bundle array
[49]	Design of a spiral finned tube HEX for closed sorption heat storage	Liquid absorption	NaOH/water	Lab-scale	The tested heat exchanger guarantees large surface area, good wetting, thin film and sufficient exposure time, but the achievable thermal power is lower than with other solutions
[50]	Design of a falling film tube bundle HEX for closed sorption heat storage	Liquid absorption	NaOH/water	Prototype—10 kW nominal power	Detailed design of the absorber HEX is crucial for the correct operation of the prototype. An estimated energy density double of those of water tank was reached
[51]	Seasonal heat storage from solar collectors	Liquid absorption	KCOOH/water	Prototype—8 kWh nominal energy and 1 kW nominal power	The system can be regenerated at temperatures of 50–60 °C, but crystallisation occurred under some conditions
[43]	Open adsorption heat storage for a single family house	Open cycle adsorption	Zeolite Na-X/air	Prototype—4 kWh energy 2 kW nominal power	The system was able to provide continuous useful effect for 6 h. Air flow rate does not influence the amount of stored energy
[44]	Open adsorption heat storage for residential applications	Open cycle adsorption	Zeolite 13X/air	Prototype—2 kW nominal power	Effective regeneration of the prototype with 120–180 °C hot air from solar collectors was possible and a useful effect for 8 h was measured

(continued)

Table 4 (continued)

References	System description	Classification	Working pair	Size	Relevant results
[52]	Open adsorption seasonal heat storage for residential applications	Open cycle adsorption	Zeolite 13XBF/air	Prototype—4 kW nominal power	The reactor was divided into 4 segments of 62.5 l each. A maximum output power of 0.9 kW per segment was achieved. About 75% of total discharge is obtained in the first 10 h.
[53]	Adsorption storage for refrigeration purposes	Open cycle adsorption	Zeolite 13XBF/air	Prototype—3 kWh nominal energy	Design of the adsorber proved to be a critical issue, especially in terms of heat transfer and vacuum tightness
[54]	Closed seasonal adsorption storage to cover domestic hot water and space heating demand of a single family house	Closed cycle adsorption	Zeolite 13X/water	Pilot demonstrator	The system proved to be reliable and able to work with a solar fraction of 83.5%. A charge boost mode operation was tested, that allowed increasing energy density. Vacuum tightness was a critical issue to be addressed during design
[55]	Adsorption seasonal heat storage for solar heat systems in residential applications realised through a segmented reactor	Closed cycle adsorption	Zeolite 13XBF/water	Pilot demonstrator	The reactor was divided into 12 segment pairs that can be adsorbed and desorbed separately. Adjustment of flow rate to keep a constant outlet power was suggested
[24]	Adsorption closed storage for heat and cold applications	Closed cycle adsorption	Zeotype FAM Z02/water	Lab-scale	The system was tested as short-term heat storage, long-term heat storage and cold storage. An increase of energy density of about 100% with respect of water under the same boundaries was measured
[56]	Evaporation systems for application in closed sorption storages	Closed cycle adsorption	–	Lab-scale	Measurements on a falling film tube bundle to be used as evaporator in sorption storages, assessing the effect of geometrical parameters, such as tube spacing and irrigation density

(continued)

Table 4 (continued)

References	System description	Classification	Working pair	Size	Relevant results
[57]	Storage for space heating applications	Chemical reaction	$SrBr_2 \cdot 6H_2O$	Lab-scale	The prototype is realised embedding the salt in an aluminium honeycomb structure to prevent agglomeration. 13 cycles were realised, showing a good cycling stability. Some design issues were evidenced, such as evapo-condenser unsteadiness, aluminium corrosion, and salt leaving the cells
[58]	Storage for space heating applications	Chemical reaction	$SrBr_2 \cdot 6H_2O$	Prototype—100 kWh nominal storage capacity	The system was tested several times, showing a stabilisation of the thermal power output after the first cycle, with an average value of 0.3 kW. Furthermore, ways to control this storage system in order to meet user demand were investigated by changing the equilibrium drop and the air flow rate
[59]	Heat storage with gas–solid reaction	Composite materials	$MnCl_2$-ENG/NH_3	Lab-scale	Expanded graphite was used to increase the thermal conductivity of the salt. High maximum power (up to 50 kW) was measured, but the need for further improvements was evidenced, in particular towards the reduction of heat losses and the increase in filling ratio
[60]	Modular open sorption pipe reactor for solar thermal energy storage	Composite materials	Vermiculite-$CaCl_2$/air	Lab-scale	The system was regenerated with temperatures as low as 80 °C. The reactor consists of an outer and an inner (perforated) cylindrical shell and the sorption material filled in between. Results indicated an increase of storable energy up to 6 times that of water

(continued)

Table 4 (continued)

References	System description	Classification	Working pair	Size	Relevant results
[61]	Short and long-term heat storage	Composite materials	Vermiculite-LiCl/water	Lab-scale	Tests were done under seasonal storage boundary conditions. Average power of 0.5 kW and energy density of 330 kWh/$kg_{material}$ were measured
[46]	Thermal battery for deep-freezing cold energy storage and heat energy storage	Composite materials	$MnCl_2$-EG/NH_3	Lab-scale	Cold and heat energy densities measured are 600 kJ/kg and 1498 kJ/kg, respectively However, the benefits are more marked for heat energy storage when compared with cold energy storage
[47]	Seasonal solar heat storage for space heating applications	Composite materials	$SrCl_2$-EG/NH_3	Prototype	The system consists of two storage units, that can work in series or in parallel. For the two modes, energy storage densities of 706 kJ/kg and 305 kJ/kg were measured. Optimization of the HEXs is indicated as the possibility to increase salt conversion and improve system performance
[62]	Open sorption heat storage	Composite materials	Na-X zeolite-$CaCl_2$/air	Lab-scale	The composite adsorbent is realised by impregnation of the zeolite with the salt. The system consists of an external reactor and a moving material bed. Steady-state power output was 1.2 kW
[63]	Heat storage for domestic hot water and space heating	Composite materials	LiCl-EG/water	Prototype—1 kWh nominal stored energy	The sorption reactor consists of 25 identical sorption bed units, which are connected by manifold tubes in parallel. Stored energy density is compared to that of a water tank and is found to be double that of water

(continued)

Table 4 (continued)

References	System description	Classification	Working pair	Size	Relevant results
[64]	Open sorption thermal storage for space heating	Composite materials	LiCl-activated alumina/air	Prototype—1 kWh nominal stored energy	The reactor is designed as a modular system. However, due to different air inlet conditions, the various modules have different performance. Useful effect (warm air at 30 °C) was supplied for 7 h
[48]	Seasonal resorption energy storage for space heating	Composite materials	$MnCl_2$-$CaCl_2$/NH_3, $MnCl_2$-$CaCl_2$-ENG/NH_3	Prototype	Different operating modes are evaluated, based on winter external conditions. Despite the typical operation with a high temperature and low temperature unit, coupling of two sets of high temperature and low temperature units and a sorption-compression mode are tested
[65]	Hybrid "solid sorption heat storage/air sourced heat pump" system for energy efficient	Composite materials	Vermiculite-$CaCl_2$/air, Vermiculite-$MgCl_2$/air. Vermiculite-LiCl-$CaCl_2$/air, Vermiculite-LiCl-$MgCl_2$/air,	Prototype—full scale domestic size	The reactor is a multilayer fixed bed. Among the composites, Vermiculite-$CaCl_2$ and Vermiculite-LiCl-$CaCl_2$ performed better. Under the best conditions, a useful effect for 1200 min, corresponding to 170 kWh could be obtained

Table 5 Boundary conditions and main results of the prototypes of sorption storages presented in literature

References	Classification	Working pair	Experimental conditions			Energy density	Max power (kW)
			T_{charge} (°C)	$T_{discharge}$ (°C)	$p_{discharge}$ (kPa)		
[66]	Liquid absorption	$CaCl_2$/water	95	47	–	380 $kWh/m^3_{material}$	–
[67]	Liquid absorption	LiBr/water	95	22	1.6	–	5.0
[42]	Liquid absorption	NaOH/water	95	25	2.6	–	1.0
[49]	Liquid absorption	NaOH/water	70	30–55	1.8	435 $kWh/m^3_{material}$	0.2
[50]	Liquid absorption	NaOH/water	70	50	5–12	74 kWh/m^3_{system}	1.2
[51]	Liquid absorption	KCOOH/water	50–70	30	4	–	0.4
[68]	Open cycle adsorption	Zeolite 4A/air	120–180	20–25	Atmospheric	130 $kWh/m^3_{material}$	–
[43]	Open cycle adsorption	Zeolite Na-X/air	180	20	Atmospheric	0.3 $kWh/kg_{material}$	2.2
[44]	Open cycle adsorption	Zeolite 13X/air	120–180	20	8	–	–
[52]	Open cycle adsorption	Zeolite 13XBF/air	180	30	Atmospheric	198 $kWh/m^3_{material}$, 109 kWh/m^3_{system}	0.9
[69]	Closed cycle adsorption	Zeolite 13XBF/water	180	65	–	95 $kWh/m^3_{material}$	–
[70]	Closed cycle adsorption	Zeolite 13X/water	175–250	70	7	20.4 kWh/m^3_{system}	–
[53]	Closed cycle adsorption	Zeolite 13XBF/water	220	20	1.2	–	0.5
[54]	Closed cycle adsorption	Zeolite 13X/water	180	25	1.8	178 $kWh/m^3_{material}$	4.6

(continued)

Table 5 (continued)

References	Classification	Working pair	Experimental conditions			Energy density	Max power (kW)
			T_{charge} (°C)	$T_{discharge}$ (°C)	$p_{discharge}$ (kPa)		
[55]	Closed cycle adsorption	Zeolite 13XBF/water	180	30	–	163 $kWh/m^3_{material}$-	0.8
[24]	Closed cycle adsorption	Zeotype FAM Z02/water	90	5–20 (cold storage) 30–45 (heat storage)	1.2	0.8 $kWh/kg_{material}$ (cold storage) 1.6 $kWh/kg_{material}$ (short-term heat storage) 2.8 $kWh/kg_{material}$ (long-term heat storage)	0.6
[56]	Closed cycle adsorption	–	–	–	–	–	–
[71]	Chemical reaction	$Na_2S{\cdot}5H_2O$	80	20	1.8	1.7 $kWh/kg_{material}$ (heat) 0.7 $kWh/kg_{material}$ (cold)-	0.5 W/kg
[72]	Chemical reaction	$MgCl_2{\cdot}6H_2O$	130	60	1.2	140 $kWh/m^3_{material}$	0.15
[58]	Chemical reaction	$SrBr_2{\cdot}6H_2O$	80	25	Atmospheric	203 $kWh/m^3_{material}$	0.8
[57]	Chemical reaction	$SrBr_2{\cdot}6H_2O$	100	20	1.5–3.2	213 $kWh/m^3_{material}$	
[59]	Composite materials	$MnCl_2$-ENG/NH_3	174	50	90	179 $kWh/m^3_{material}$	9.9
[73]	Composite materials	$BaCl_2$-ENG/NH_3	60–70	4	–	114 $kWh/kg_{material}$	5.0
[60]	Composite materials	Vermiculite-$CaCl_2$/air	80	20	Atmospheric	290 $kWh/m^3_{material}$	0.6
[61]	Composite materials	Vermiculite-LiCl/water	75–90	35	1.2	2330 $kWh/kg_{material}$	

(continued)

Table 5 (continued)

References	Classification	Working pair	Experimental conditions			Energy density	Max power (kW)
			T_{charge} (°C)	$T_{discharge}$ (°C)	$p_{discharge}$ (kPa)		
[46]	Composite materials	$MnCl_2$-EG/NH_3	150	112 °C (heat storage) −10 °C (cold storage)	–	167 kWh/$kg_{material}$ (cold storage) 417 kWh/$kg_{material}$ (heat storage)	–
[47]	Composite materials	$SrCl_2$-EG/NH_3	90	45	–	196 kWh/$kg_{material}$ (parallel operation) 85 kWh/$kg_{material}$ (series operation)	643 W/$kg_{material}$
[62]	Composite materials	Na-X zeolite-$CaCl_2$/air	180	30	1.5	260 kWh/$m^3_{material}$	1.5
[63]	Composite materials	LiCl-EG/water	85	40	1.5	65 kWh/$m^3_{material}$	11
[64]	Composite materials	LiCl-activated alumina/air	110	20	Atmospheric	190 kWh/$m^3_{material}$	0.13
[48]	Composite materials	$MnCl_2$-$CaCl_2$/NH_3, $MnCl_2$-$CaCl_2$-ENG/NH_3	50	30	–	1047 kJ/$kg_{material}$l	402 W/$kg_{material}$
[65]	Composite materials	Vermiculite-$CaCl_2$/air, Vermiculite-$MgCl_2$/air. Vermiculite-LiCl-$CaCl_2$/air Vermiculite-LiCl-$MgCl_2$/air	85	30	Atmospheric	170 kWh/$m^3_{material}$	

in [46, 47] are used as 'sorption thermal battery' for cold and heat energy storage: based on user request (i.e. cold or heat) connection to the evaporator or to the sorption bed of the thermal battery can be selected, even to achieve contemporary heat and freezing useful effect. The prototype reported in [48] is a resorption thermal battery that can be operated in different modes according to the external ambient temperature.

A common outcome of many experimentations is the need to address design issues, either to increase the achievable storage capacity or, in the case of the prototypes containing salts, to avoid salt leakage and increase the thermal conductivity of the system. It is however possible to foresee that, thanks also to the increasing number of projects involving sorption storages, a technological growth will be achieved within the next years.

References

1. McBain JW (1909) XCIX. The mechanism of the adsorption ("sorption") of hydrogen by carbon. Philos Mag Ser 6 18:916–935. https://doi.org/10.1080/14786441208636769
2. Cabeza LF, Solé A, Barreneche C (2017) Review on sorption materials and technologies for heat pumps and thermal energy storage. Renew Energy 110:3–39. https://doi.org/10.1016/j.renene.2016.09.059
3. Ding Y, Riffat SB (2013) Thermochemical energy storage technologies for building applications: a state-of-the-art review. Int J Low-Carbon Technol 8:106–116
4. Lefebvre D, Tezel FH (2017) A review of energy storage technologies with a focus on adsorption thermal energy storage processes for heating applications. Renew Sustain Energy Rev 67:116–125
5. Yu N, Wang RZ, Wang LW (2013) Sorption thermal storage for solar energy. Prog Energy Combust Sci 39:489–514. https://doi.org/10.1016/j.pecs.2013.05.004
6. Scapino L, Zondag HA, Van Bael J et al (2017) Sorption heat storage for long-term low-temperature applications: a review on the advancements at material and prototype scale. Appl Energy 190:920–948. https://doi.org/10.1016/j.apenergy.2016.12.148
7. Chen JF, Dai YJ, Wang RZ (2017) Experimental and analytical study on an air-cooled single effect LiBr-H_2O absorption chiller driven by evacuated glass tube solar collector for cooling application in residential buildings. Sol Energy 151:110–118. https://doi.org/10.1016/j.solener.2017.05.029
8. N 'Tsoukpoe KE, Perier-Muzet M, Le Pierrès N et al (2014) Thermodynamic study of a LiBr–H_2O absorption process for solar heat storage with crystallisation of the solution. Sol Energy 104:2–15. https://doi.org/10.1016/j.solener.2013.07.024
9. Weber R, Dorer V (2008) Long-term heat storage with NaOH. Vacuum 82:708–716. https://doi.org/10.1016/J.VACUUM.2007.10.018
10. Liu H, N 'Tsoukpoe KE, Le Pierrès N, Luo L (2011) Evaluation of a seasonal storage system of solar energy for house heating using different absorption couples. Energy Convers Manag 52:2427–2436. https://doi.org/10.1016/J.ENCONMAN.2010.12.049
11. Ibrahim NI, Al-Sulaiman FA, Ani FN (2018) Solar absorption systems with integrated absorption energy storage—a review. Renew Sustain Energy Rev 82:1602–1610. https://doi.org/10.1016/J.RSER.2017.07.005
12. Siddiqui MU, Said SAM (2015) A review of solar powered absorption systems. Renew Sustain Energy Rev 42:93–115. https://doi.org/10.1016/J.RSER.2014.10.014

13. Lehmann C, Beckert S, Nonnen T et al (2017) Water loading lift and heat storage density prediction of adsorption heat storage systems using Dubinin-Polanyi theory—comparison with experimental results. Appl Energy 207:274–282. https://doi.org/10.1016/J.APENERGY.2017.07.008
14. Henninger SK, Schmidt FP, Henning H-M (2010) Water adsorption characteristics of novel materials for heat transformation applications. Appl Therm Eng 30:1692–1702. https://doi.org/10.1016/J.APPLTHERMALENG.2010.03.028
15. Deshmukh H, Maiya MP, Srinivasa Murthy S (2017) Study of sorption based energy storage system with silica gel for heating application. Appl Therm Eng 111:1640–1646. https://doi.org/10.1016/J.APPLTHERMALENG.2016.07.069
16. Gordeeva LG, Solovyeva MV, Aristov YI (2016) NH_2-MIL-125 as a promising material for adsorptive heat transformation and storage. Energy 100:18–24. https://doi.org/10.1016/J.ENERGY.2016.01.034
17. Database of zeolite structures. http://www.iza-structure.org/databases/. Accessed 28 Nov 2017
18. Alby D, Salles F, Fullenwarth J, Zajac J (2017) On the use of metal cation-exchanged zeolites in sorption thermochemical storage: some practical aspects in reference to the mechanism of water vapor adsorption. Sol Energy Mater Sol Cells. https://doi.org/10.1016/j.solmat.2017.11.020
19. Aprea P, de Gennaro B, Gargiulo N et al (2016) Sr-, Zn- and Cd-exchanged zeolitic materials as water vapor adsorbents for thermal energy storage applications. Appl Therm Eng 106:1217–1224. https://doi.org/10.1016/J.APPLTHERMALENG.2016.06.066
20. Ng E-P, Mintova S (2008) Nanoporous materials with enhanced hydrophilicity and high water sorption capacity. Microporous Mesoporous Mater 114:1–26. https://doi.org/10.1016/j.micromeso.2007.12.022
21. Brancato V, Frazzica A (2018) Characterisation and comparative analysis of zeotype water adsorbents for heat transformation applications. Sol Energy Mater Sol Cells 180:91–102. https://doi.org/10.1016/J.SOLMAT.2018.02.035
22. Sing KSW, Everett DH, Haul RAW et al (1985) Reporting physisorption data for gas/solid systems—with special reference to the determination of surface area and porosity. Pure Appl Chem 57:603–619
23. Aristov YI (2014) Concept of adsorbent optimal for adsorptive cooling/heating. Appl Therm Eng 72:166–175. https://doi.org/10.1016/J.APPLTHERMALENG.2014.04.077
24. Palomba V, Vasta S, Freni A (2017) Experimental testing of AQSOA FAM Z02/water adsorption system for heat and cold storage. Appl Therm Eng 124:967–974. https://doi.org/10.1016/J.APPLTHERMALENG.2017.06.085
25. Bales C, Gantenbein P, Jaenig D et al (2008) Laboratory tests of chemical reactions and prototype sorption storage units. A report of IEA solar heating and cooling programme—task 32: advanced storage concepts for solar and low energy buildings
26. Chi CW, Wasan DT (1969) Measuring the equilibrium pressure of supported and unsupported adsorbents. Ind Eng Chem Fundam 8:816–818. https://doi.org/10.1021/i160032a036
27. Heiti RV, Thodos G (1986) Energy release in the dehumidification of air using a bed of $CaCl_2$-impregnated Celite. Ind Eng Chem Fundam 25:768–771. https://doi.org/10.1021/i100024a048
28. Casey SP, Elvins J, Riffat S, Robinson A (2014) Salt impregnated desiccant matrices for "open" thermochemical energy storage—selection, synthesis and characterisation of candidate materials. Energy Build 84:412–425. https://doi.org/10.1016/J.ENBUILD.2014.08.028
29. Druske M-M, Fopah-Lele A, Korhammer K et al (2014) Developed materials for thermal energy storage: synthesis and characterization. Energy Proc 61:96–99. https://doi.org/10.1016/J.EGYPRO.2014.11.915
30. Liu H, Nagano K, Togawa J (2015) A composite material made of mesoporous siliceous shale impregnated with lithium chloride for an open sorption thermal energy storage system. Sol Energy 111:186–200. https://doi.org/10.1016/J.SOLENER.2014.10.044

31. Opel O, Rammelberg HU, Gerard M, Ruck WKL (2011) Thermochemical storage materials research—TGA/DSC-hydration studies. In: 1st International conference for sustainable energy storage
32. Yu N, Wang RZ, Lu ZS, Wang LW (2015) Study on consolidated composite sorbents impregnated with LiCl for thermal energy storage. Int J Heat Mass Transf 84:660–670. https://doi.org/10.1016/J.IJHEATMASSTRANSFER.2015.01.065
33. Ponomarenko IV, Glaznev IS, Gubar AV et al (2010) Synthesis and water sorption properties of a new composite "$CaCl_2$ confined into SBA-15 pores". Microporous Mesoporous Mater 129:243–250. https://doi.org/10.1016/J.MICROMESO.2009.09.023
34. Hongois S, Kuznik F, Stevens P, Roux J-J (2011) Development and characterisation of a new $MgSO_4$—zeolite composite for long-term thermal energy storage. Sol Energy Mater Sol Cells 95:1831–1837. https://doi.org/10.1016/j.solmat.2011.01.050
35. Jänchen J, Ackermann D, Stach H, Brösicke W (2004) Studies of the water adsorption on Zeolites and modified mesoporous materials for seasonal storage of solar heat. Sol Energy 76:339–344. https://doi.org/10.1016/J.SOLENER.2003.07.036
36. Jänchen J, Ackermann D, Weiler E et al (2005) Calorimetric investigation on zeolites, $AlPO_4$'s and $CaCl_2$ impregnated attapulgite for thermochemical storage of heat. Thermochim Acta 434:37–41. https://doi.org/10.1016/J.TCA.2005.01.009
37. Posern K, Kaps C (2010) Calorimetric studies of thermochemical heat storage materials based on mixtures of $MgSO_4$ and $MgCl_2$. Thermochim Acta 502:73–76. https://doi.org/10.1016/J.TCA.2010.02.009
38. Ristić A, Maučec D, Henninger SK, Kaučič V (2012) New two-component water sorbent $CaCl_2$-FeKIL2 for solar thermal energy storage. Microporous Mesoporous Mater 164:266–272. https://doi.org/10.1016/J.MICROMESO.2012.06.054
39. Korhammer K, Druske M-M, Fopah-Lele A et al (2016) Sorption and thermal characterization of composite materials based on chlorides for thermal energy storage. Appl Energy 162:1462–1472. https://doi.org/10.1016/j.apenergy.2015.08.037
40. Grekova A, Gordeeva L, Aristov Y (2016) Composite sorbents "Li/Ca halogenides inside multi-wall carbon nano-tubes" for thermal energy storage. Sol Energy Mater Sol Cells 155:176–183. https://doi.org/10.1016/j.solmat.2016.06.006
41. Grekova AD, Gordeeva LG, Aristov YI (2017) Composite "LiCl/vermiculite" as advanced water sorbent for thermal energy storage. Appl Therm Eng 124:1401–1408. https://doi.org/10.1016/J.APPLTHERMALENG.2017.06.122
42. Fumey B, Weber R, Gantenbein P et al (2015) Operation results of a closed sorption heat storage prototype. Energy Proc 73:324–330. https://doi.org/10.1016/j.egypro.2015.07.698
43. Johannes K, Kuznik F, Hubert J-L et al (2015) Design and characterisation of a high powered energy dense zeolite thermal energy storage system for buildings. Appl Energy 159:80–86. https://doi.org/10.1016/J.APENERGY.2015.08.109
44. Tatsidjodoung P, Le Pierrès N, Heintz J et al (2016) Experimental and numerical investigations of a zeolite 13X/water reactor for solar heat storage in buildings. Energy Convers Manag 108:488–500. https://doi.org/10.1016/J.ENCONMAN.2015.11.011
45. Gaeini M, Javed MR, Ouwerkerk H et al (2017) Realization of a 4 kW thermochemical segmented reactor in household scale for seasonal heat storage. Energy Proc 135:105–114. https://doi.org/10.1016/j.egypro.2017.09.491
46. Li TX, Wu S, Yan T et al (2016) A novel solid–gas thermochemical multilevel sorption thermal battery for cascaded solar thermal energy storage. Appl Energy 161:1–10. https://doi.org/10.1016/j.apenergy.2015.09.084
47. Li TX, Wu S, Yan T et al (2017) Experimental investigation on a dual-mode thermochemical sorption energy storage system. Energy 140:383–394. https://doi.org/10.1016/j.energy.2017.08.073
48. Jiang L, Wang RZ, Wang LW, Roskilly AP (2017) Investigation on an innovative resorption system for seasonal thermal energy storage. Energy Convers Manag 149:129–139. https://doi.org/10.1016/j.enconman.2017.07.018

49. Fumey B, Weber R, Baldini L (2017) Liquid sorption heat storage—a proof of concept based on lab measurements with a novel spiral fined heat and mass exchanger design. Appl Energy 200:215–225. https://doi.org/10.1016/j.apenergy.2017.05.056
50. Daguenet-Frick X, Gantenbein P, Müller J et al (2017) Seasonal thermochemical energy storage: comparison of the experimental results with the modelling of the falling film tube bundle heat and mass exchanger unit. Renew Energy 110:162–173. https://doi.org/10.1016/j.renene.2016.10.005
51. Le Pierrès N, Huaylla F, Stutz B, Perraud J (2017) Long-term solar heat storage process by absorption with the KCOOH/H_2O couple: experimental investigation. Energy 141:1313–1323. https://doi.org/10.1016/j.energy.2017.10.111
52. Gaeini M, Rouws AL, Salari JWO et al (2018) Characterization of microencapsulated and impregnated porous host materials based on calcium chloride for thermochemical energy storage. Appl Energy 212:1165–1177. https://doi.org/10.1016/j.apenergy.2017.12.131
53. Semprini S, Asenbeck S, Kerskes H, Drück H (2017) Experimental and numerical investigations of an adsorption water-zeolite heat storage for refrigeration applications. Energy Proc 135:513–521. https://doi.org/10.1016/j.egypro.2017.09.492
54. Köll R, van Helden W, Engel G et al (2017) An experimental investigation of a realistic-scale seasonal solar adsorption storage system for buildings. Sol Energy 155:388–397. https://doi.org/10.1016/j.solener.2017.06.043
55. Weber R, Asenbeck S, Kerskes H, Drück H (2016) SolSpaces—testing and performance analysis of a segmented sorption store for solar thermal space heating. Energy Proc 91:250–258. https://doi.org/10.1016/j.egypro.2016.06.214
56. Dang BN, Van Helden W, Luke A (2017) Investigation of water evaporation for closed sorption storage systems. Energy Proc 135:504–512. https://doi.org/10.1016/j.egypro.2017.09.493
57. Fopah-Lele A, Rohde C, Neumann K et al (2016) Lab-scale experiment of a closed thermochemical heat storage system including honeycomb heat exchanger. Energy 114:225–238. https://doi.org/10.1016/j.energy.2016.08.009
58. Michel B, Mazet N, Neveu P (2014) Experimental investigation of an innovative thermochemical process operating with a hydrate salt and moist air for thermal storage of solar energy: global performance. Appl Energy 129:177–186. https://doi.org/10.1016/j.apenergy.2014.04.073
59. Yan T, Wang RZ, Li TX (2018) Experimental investigation on thermochemical heat storage using manganese chloride/ammonia. Energy 143:562–574. https://doi.org/10.1016/j.energy.2017.11.030
60. Aydin D, Casey SP, Chen X, Riffat S (2016) Novel "open-sorption pipe" reactor for solar thermal energy storage. Energy Convers Manag 121:321–334. https://doi.org/10.1016/j.enconman.2016.05.045
61. Palomba V, Gordeeva L, Brancato V et al (2017) Experimental characterization of a lab-scale adsorption thermal storage based on the LiCl/vermiculite composite sorbent. In: International sorption heat pump conference. Tokyo
62. Nonnen T, Beckert S, Gleichmann K et al (2016) A thermochemical long-term heat storage system based on a salt/zeolite composite. Chem Eng Technol 39:2427–2434. https://doi.org/10.1002/ceat.201600301
63. Zhao YJ, Wang RZ, Li TX, Nomura Y (2016) Investigation of a 10 kWh sorption heat storage device for effective utilization of low-grade thermal energy. Energy 113:739–747. https://doi.org/10.1016/j.energy.2016.07.100
64. Zhang YN, Wang RZ, Li TX (2017) Experimental investigation on an open sorption thermal storage system for space heating. Energy 141:2421–2433. https://doi.org/10.1016/j.energy.2017.12.003
65. Aydin D, Casey SP, Chen X, Riffat S (2018) Numerical and experimental analysis of a novel heat pump driven sorption storage heater. Appl Energy 211:954–974. https://doi.org/10.1016/j.apenergy.2017.11.102

66. Quinnell JA, Davidson JH (2012) Mass transfer during sensible charging of a hybrid absorption/sensible storage tank. Energy Proc 30:353–361. https://doi.org/10.1016/J.EGYPRO.2012.11.042
67. N 'Tsoukpoe KE, Le Pierrès N, Luo L (2012) Experimentation of a LiBr-H_2O absorption process for long term solar thermal storage. Energy Proc 30:331–341. https://doi.org/10.1016/J.EGYPRO.2012.11.039
68. Mette B, Kerskes H, Drück H, Müller-Steinhagen H (2013) New highly efficient regeneration process for thermochemical energy storage. Appl Energy 109:352–359. https://doi.org/10.1016/j.apenergy.2013.01.087
69. Weber R, Kerskes H, Drück H (2014) Development of a combined hot water and sorption store for solar thermal systems. Energy Proc 48:464–473. https://doi.org/10.1016/j.egypro.2014.02.055
70. Schreiber H, Lanzerath F, Reinert C et al (2016) Heat lost or stored: Experimental analysis of adsorption thermal energy storage. Appl Therm Eng 106:981–991. https://doi.org/10.1016/j.applthermaleng.2016.06.058
71. De Boer R, Haije WG, Veldhuis JBJ, Smeding SF (2004) Solid-sorption cooling with integrated thermal storage: the SWEAT prototype. In: Heat powered cycles, HPC 2004. Larnaca
72. Zondag H, Kikkert B, Smeding S et al (2013) Prototype thermochemical heat storage with open reactor system. Appl Energy 109:360–365. https://doi.org/10.1016/j.apenergy.2013.01.082
73. Stitou D, Mazet N, Mauran S (2012) Experimental investigation of a solid/gas thermochemical storage process for solar air-conditioning. Energy 41:261–270. https://doi.org/10.1016/j.energy.2011.07.029

Materials Selection for Thermal Energy Storage Applications—Case Studies

Mercè Segarra, Camila Barreneche, Alejandro Calderón and Ana Inés Fernández

Abstract To accomplish their duties, engineers need information, and this must be easily accessible, so they can readily choose among all the possibilities they can find to fit to a specific application. To build a thermal energy storage system, engineers always wonder which the best storage material they can find. The answer always depends on several factors. In the present chapter, the materials selection methodology is introduced to proceed for an optimal material selection for a certain application in thermal energy storage systems. Several case studies using this methodology are explained for different thermal energy storage applications: long term and short term sensible heat thermal energy storage, and phase change materials (PCM) for latent heat thermal energy storage for low temperature applications (buildings sector), and high temperature applications (concentrated solar power plants or CSP).

1 Introduction

The innovation regarding materials selection and new materials development for thermal energy storage (TES) applications is one of the main challenges to enhance the deployment of these technologies in energy productive sectors such as concentrated solar power plants (CSP), or in the improvement of energy efficiency in buildings, for example.

M. Segarra · C. Barreneche (✉) · A. Calderón · A. I. Fernández
Department of Materials Science and Physical Chemistry,
Universitat de Barcelona, Martí I Franquès 1, 08028 Barcelona, Spain
e-mail: c.barreneche@ub.edu

C. Barreneche
School of Chemical Engineering, BCES (Birmingham Centre of Energy Storage), University of Birmingham, Birmingham B15 2TT, UK

A. Frazzica and L. F. Cabeza (eds.), *Recent Advancements in Materials and Systems for Thermal Energy Storage*, Green Energy and Technology,
https://doi.org/10.1007/978-3-319-96640-3_5

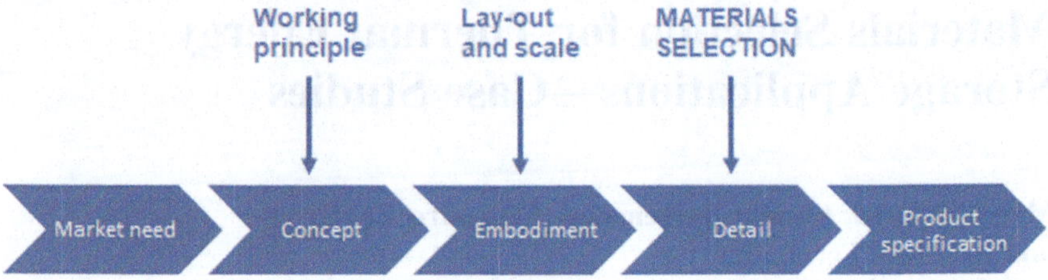

Fig. 1 Materials selection within system design process

The usual path followed to develop technological improvements is based on system design. Once the complete system is designed, companies and/or researchers look for the available materials that best fit the specified working conditions of the designed system. However, it is very common to find out that there is no material for such application that meets all the requirements of the system, or that its price or availability makes it unfeasible. That is why materials selection plays a key role as an initial step to design the system [1].

System design (Fig. 1) starts by establishing a set of requirements that could be met by different ways or concepts, from which only one is selected and developed (design embodiment). Once this chosen concept is refined, the process leads to a detailed design in which materials, that must have been accompanying the whole process, can be finally optimally chosen.

To follow this materials-based design procedure, it is important to identify the parameters and attributes for materials to meet the requirements of the system in order to perform the selection of an optimal material in a proper way.

The evaluation of the material performance under the working boundary conditions is crucial during a design process. This performance for a specific application is determined by the material index [2], which is a mathematical equation that relates some relevant properties of materials. This metric facilitates the material selection of the most suitable material to be applied for such application. This parameter is calculated based on several considerations listed below:

- **Function**: it describes what the component/product principal does in the system.
- **Constraints**: there are some mandatory requirements the material must fulfil, such as temperature range, fixed dimensions or withstanding loads without failure. Constrains are normally expressed as boundary parameters of the material.
- **Objective**: stands for the performance of the application, from which an extreme value is sought. It is a key issue to proceed with the selection procedure and it is usually expressed as something that should be maximized or minimized, for example to be low cost, light weighted or a combination of them.
- **Free variables**: these are those properties/parameters that can be changed and are not specifically defined in the design.

2 Case Studies of Sensible Heat Energy Storage

The ability to store energy as sensible heat for a given material strongly depends on the value of its energy density, that is the heat capacity per unit volume or ρC_p, without phase change in the temperature range of the storage process. On the other hand, for a material to be useful in a TES application, it must be inexpensive and have good thermal conductivity. Two scenarios have been proposed by Fernández et al. [3] to store thermal energy as sensible heat: long term and short term. In these case studies, the working temperature range of 150–200 °C has been considered.

2.1 Long-Term Energy Storage

Consider the case of long-term energy storage, in which materials must store the maximum amount of heat as possible.

Materials must present a high energy density, a high thermal conductivity (to decrease the thermal stratification within the bulk material storage), they must withstand temperatures higher than 150–200 °C with no chemical or physical change, and above all, they must be low cost. Then, which is the best material candidate to be used to store sensible heat for a long term? Boundary conditions for this case study are summarized in Table 1.

If we consider a mass, m, of solid material, heated through a temperature interval ΔT, the thermal energy stored by the material per unit volume and cost is:

$$\frac{Q}{C} = \frac{C_p}{C_m} \Delta T \tag{1}$$

where Q is the total energy stored (in J), C is the cost of the material (in €), C_p is the specific heat capacity of the solid (in kJ/kg K), and C_m is the material cost per unit mass (in €/kg). From this equation, material index (M_1) can be extracted, being:

$$M_1 = \frac{C_p}{C_m} \tag{2}$$

Table 1 Boundary conditions for material selection: long-term TES as sensible heat

Function	Long term—thermal energy storage
Constraints	High energy density (or heat capacity per unit volume) Maximum service temperature higher than 150 °C Good thermal conductivity (above 1.0 W/(m K))
Objective	Maximize the energy storage per unit of material cost Maximize thermal conductivity
Free variables	Dimensions Choice of material

Using the CES Selector [4], in Fig. 2 C_p is plotted versus C_m. In this Figure, several common materials are presented, which fulfil the requirements reported in Table 1. Materials that maximize (M_1) directly maximize the thermal energy stored per unit volume and cost. These can be found at the upper left part of Fig. 2, such as halite (NaCl), several concrete types referred as normal Portland, super sulphate cement, high alumina cement, etc. Straight lines represent a family of parallel lines with a slope of 1, corresponding each line to a different value of M_1. All the materials that lie on a line of constant M_1 (indicated by a red arrow in Fig. 2) perform equally well.

Short-term thermal energy storage is required in solar cooling for instance. In this case, materials must store the highest amount of energy for a given temperature rise and time, to further transfer it to another medium. That is, materials must store heat and easily release it, what can be translated to high thermal diffusivity. Thus, boundary conditions for this case study can be summarized in Table 2.

In this case, two objectives are defined, which lead to 2 different material indices. In the first one, we consider a mass, m, of solid material, heated through a temperature interval ΔT, for a given time, t. The thermal energy stored by the material per unit area is:

$$\frac{Q}{A} = \frac{3}{\sqrt{2}} t^{1/2} \frac{\lambda}{a^{1/2}} \Delta T \tag{3}$$

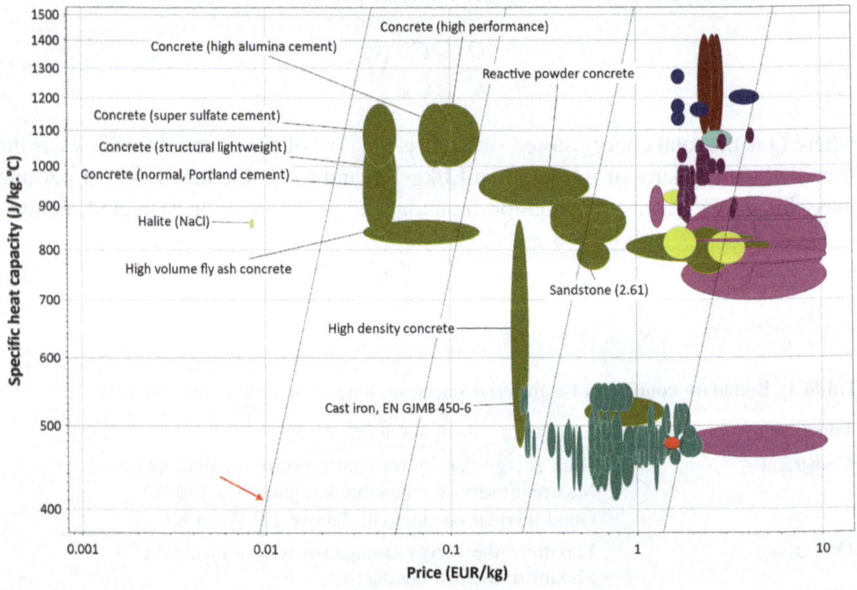

Fig. 2 Materials to be used as sensible heat storage material for long term. Chart from CES Selector 2018 [4]

Table 2 Boundary conditions for material selection: short-term TES as sensible heat

Function	Short term—thermal energy storage
Constraints	High energy density (or heat capacity per unit volume) Maximum service temperature higher than 150 °C Given temperature rise and time
Objective	Maximize the energy storage for a given temperature rise and time Minimize cost
Free Variables	Material thickness Choice of material

where λ is the thermal conductivity of the material (in W/m K), and a is its thermal diffusivity (in m^2/s). From this equation, material index (M_2) can be extracted, being:

$$M_2 = \frac{\lambda}{a^{1/2}} \tag{4}$$

Figure 3 shows those materials that fulfil the requirements previously defined in Table 2 and maximize M_2 (upper left part of the graph), such as some technical ceramics (yellow circles), or some copper alloys (red circles). Each parallel line with a slope of 0.5 (indicated with a red arrow) corresponds to a different value of M_2. All the materials lying on the same line perform equally well.

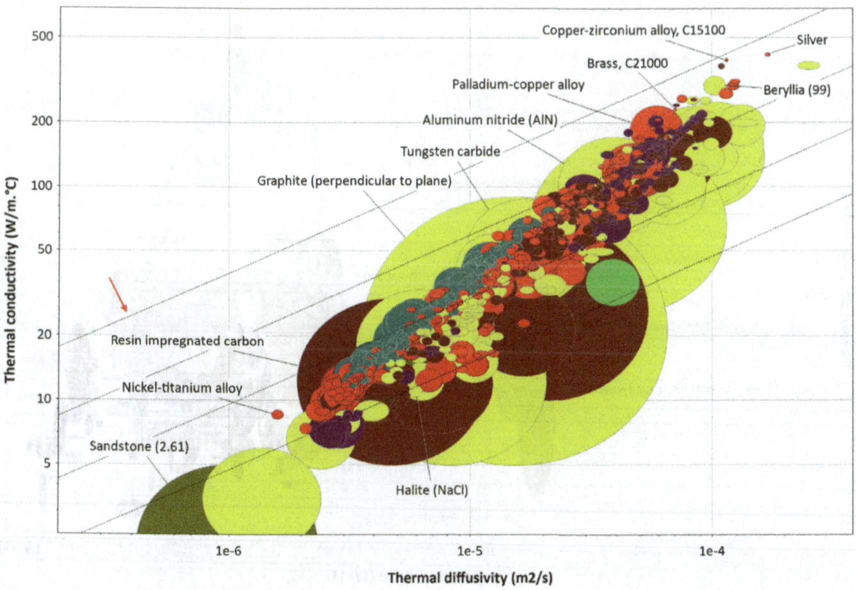

Fig. 3 Materials to be used as sensible heat storage material for short term maximizing heat storage. Chart from CES Selector 2018 [4]

Cost is the objective in the second material index. Considering a mass, m, of solid material, heated through a temperature interval ΔT, for a given time, t, in the transient state, and considering that the thickness of the material, x, is related to the time needed for heat to flow across it by its thermal diffusivity, the cost of material per unit area becomes:

$$\frac{C}{A} = \sqrt{2}ta^{1/2}\rho C_{\mathrm{m}} \tag{5}$$

where ρ is the material density (in kg/m^3).

From this equation, material index (M_3) can be extracted, being:

$$M_3 = \frac{1}{a^{1/2}\rho C_{\mathrm{m}}} \tag{6}$$

Figure 4 shows those materials that maximize M_3 (bottom left part of the graph) and fulfil the requirements previously defined in Table 2. Each parallel line with a slope of-2 (indicated with a red arrow) corresponds to a different value of M_3.

Materials performing the best should store the highest amount of energy and be the cheapest. As there is not a unique class of material that optimizes both objectives, the final material selection involves a compromise between both. Figure 5 shows the two material indices previously deduced:

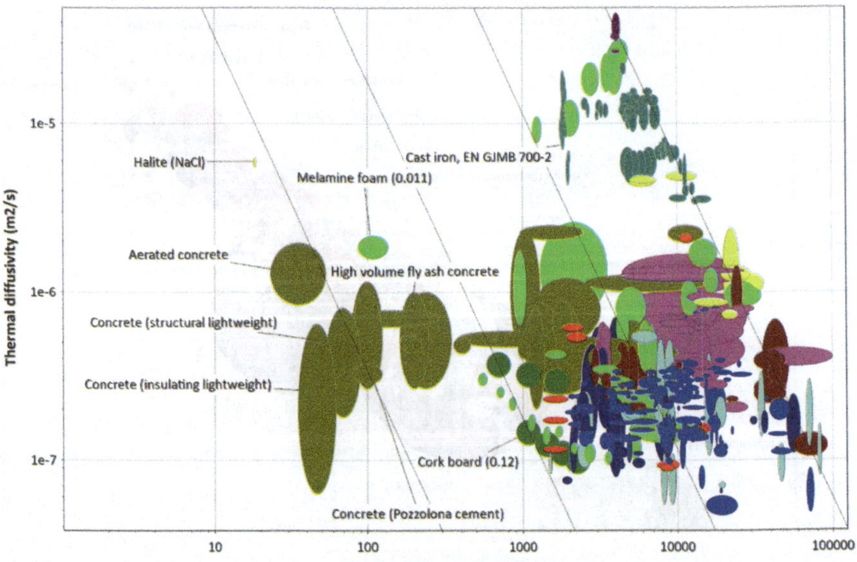

Fig. 4 Materials to be used as sensible heat storage material for short term minimizing the cost. Chart from CES Selector 2018 [4]

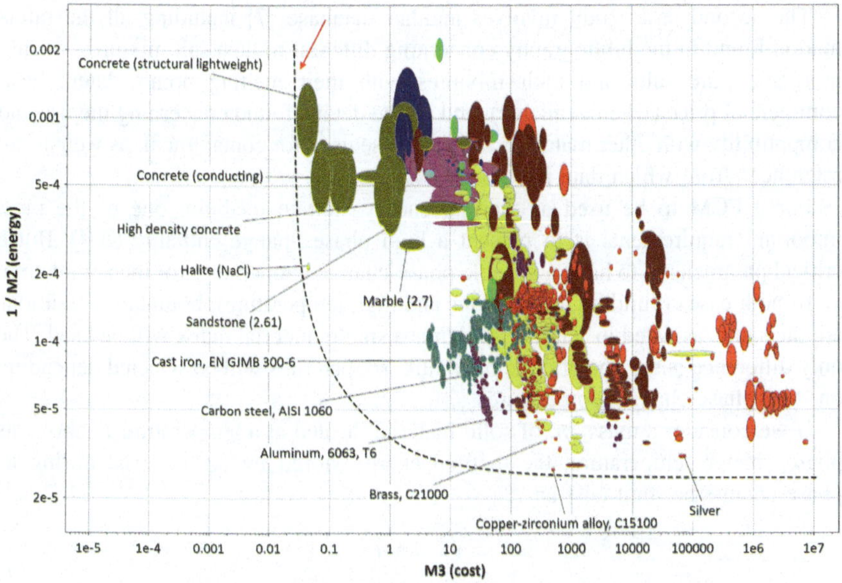

Fig. 5 Materials to be used as sensible heat storage material for short-term maximizing heat storage and release and minimizing cost. Chart from CES Selector 2018 [4]

- M_2: high energy storage and easy release
- M_3: low cost.

For trade-off curves, the convention is to invert the material indices if necessary so that the objectives are optimized when the indices are minimized. Thus, Fig. 5 is a plot of the inverse of M_2 versus the second M_3. Materials performing the best (that optimize both objectives, i.e. high energy storage and easy heat release at minimum cost) would be in the bottom left corner of the chart, or near the trade-off surface. That results in several types of concrete, halite (NaCl), cast irons and some steels, some aluminum alloys, and some copper alloys.

3 Case Studies of Latent Heat Energy Storage

To create the following case studies, different databases have been used, previously developed with CES Selector [4]. The first case study is about phase change materials (PCM) selection for building applications, as proposed by Barreneche et al. [5]. The used database includes, for each material (commercial and standard substances), at least its latent heat of fusion and its melting temperature but also, thermal conductivity, density, etc. when available, based on the information collected from bibliography [6].

The second case study involves another database [7] including all the information found in the bibliography concerning different molten salt mixtures, mainly inorganic pure salts and their mixtures with their melting points, latent heat, commercial price (when available), and all the details found concerning the use and compatibility with other materials such as those used for containment, as well as the references from which data have been extracted.

For a PCM to be used as thermal energy storage medium, one of the most important requirements is to present a high phase change enthalpy (ΔH). But a critical parameter is to have a suitable phase-change temperature for the application.

In both cases (building applications and high temperature), boundary conditions are the same, as listed in Table 3, and thus a single material index will be used. The only difference is the operation temperature, so materials will be selected depending on their phase-change temperatures.

If we consider a mass, m, of solid material, heated at a temperature T, above its phase change temperature, the thermal energy stored by the material during its phase change per unit cost is:

$$\frac{Q}{C} = \frac{\Delta H}{C_{\mathrm{m}}} \tag{7}$$

where Q is the total energy stored (in J), C is the cost of the material (in €), ΔH is the latent heat of fusion (in J/kg), and C_{m} is the material cost per unit mass (in €/kg). From this equation, material index (M_4) can be extracted, being:

$$M_4 = \frac{\Delta H}{C_{\mathrm{m}}} \tag{8}$$

Nevertheless, before reaching its melting temperature, the material stores energy while being heated, which is equivalent to what can be stored as a sensible heat. Therefore, for the whole process, the amount of energy due to phase change is increased by the amount of energy absorbed by the material when increasing its temperature up to its melting point.

We consider a mass, m, of solid, heated from room temperature up to its melting temperature. The total amount of stored energy per unit cost is given by the following equation:

Table 3 Boundary conditions for material selection of PCMs

Function	Thermal energy storage as latent heat
Constraints	Suitable phase-change temperature for the application High energy density (or heat capacity per unit volume) Maximum service temperature higher than operation temperature
Objective	Maximize the phase change enthalpy (ΔH) per unit of material cost
Free Variables	Dimensions Choice of material

$$\frac{Q_{\text{total}}}{C} = \frac{\Delta H}{C_{\text{m}}} + \frac{C_{\text{p}}\Delta T}{C_{\text{m}}} = \frac{\Delta H + C_{\text{p}}\Delta T}{C_{\text{m}}} \tag{9}$$

Next sections describe the selection of the material performing the best as a PCM for each range of temperatures.

3.1 Low Temperature PCM—Building Application

As information about cost of low temperature PCMs is not available, the selection of materials has been made just considering that the best material will be that presenting the higher value for phase change enthalpy. Figure 6 shows the latent heat of fusion versus melting temperature for all the PCM available in the database to improve the thermal comfort in building indoors. The working temperature range of this application is between 18 and 32 °C.

The best materials to be used in these systems can be found the upper side of the graph: organic paraffin wax, fatty acids and some salt hydrates like nitrate or chloride mixture salts.

On the other hand, in domestic hot water (DHW) systems the working temperature is between 45 and 65 °C. Figure 7 shows all PCM available in the database for this range of temperatures.

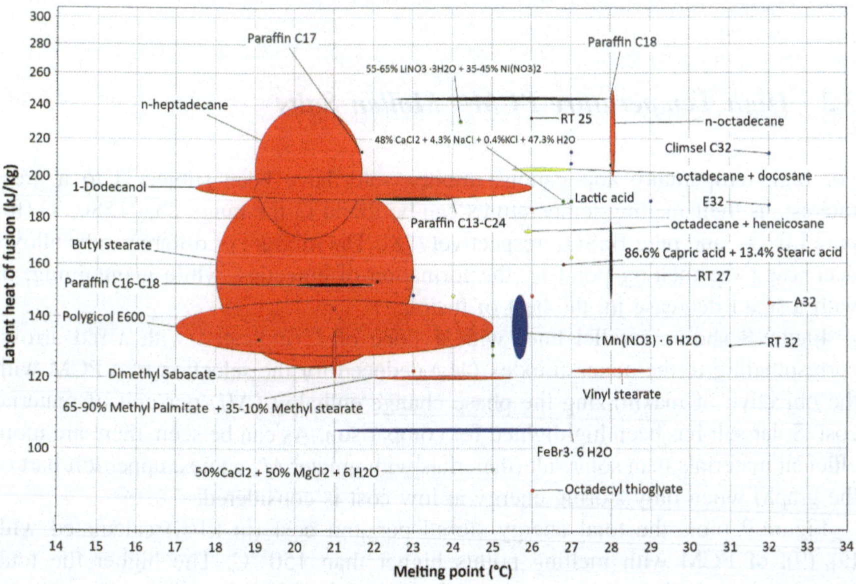

Fig. 6 Materials to be used as PCM to improve thermal comfort in buildings indoors. Chart from CES Selector 2018 [4]. Adapted from [5]

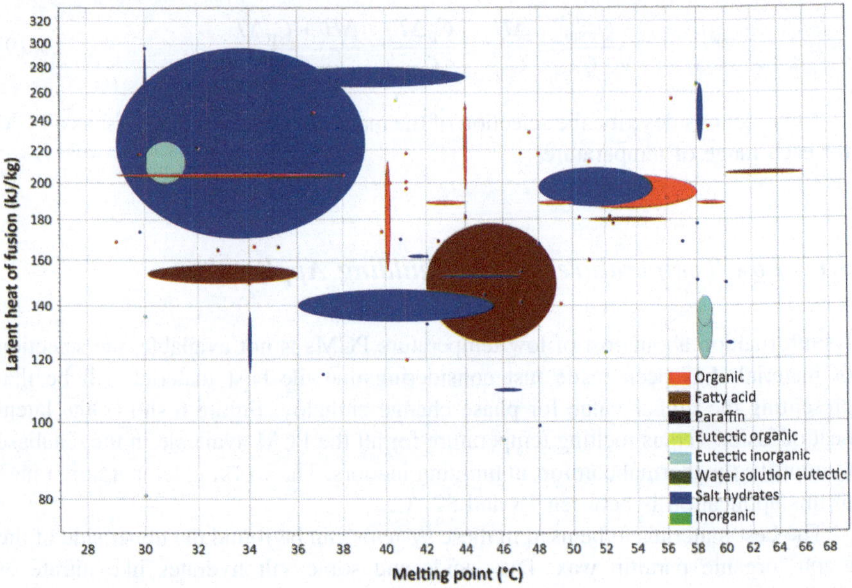

Fig. 7 Materials to be used as PCM for domestic hot water in buildings. Chart from CES Selector 2018 [4]. Adapted from [5]

To improve the energy efficiency in buildings DHW storage, the best PCM candidates are found in the upper side of the graph and correspond to salt hydrates.

3.2 *High Temperature PCM—Molten Salts*

For high temperature applications, molten salts have been subjected to a great interest, as their melting temperatures can be found in the range 253–1580 °C (for pure $LiNO_3$, and pure $BaSO_4$, respectively [8]. The mixture of different salts allows decreasing the melting point by the formation of eutectics, while maintaining, or with a small decrease in, the heat of fusion.

Figure 8 shows parallel lines with a slope of 1 (indicated with a red arrow) corresponding to the material index (M_4) deduced for the selection of a PCM with the objective of maximizing the phase change enthalpy (ΔH) per unit of material cost. Solar salt has been highlighted for comparison. As can be seen, there are more efficient materials than solar salt (materials with greater M_4 values, upper left part of the graph) when only storing energy at low cost is considered.

Figure 9 plots the total energy stored per unit cost (in kJ/€) calculated with Eq. (9), of PCM with melting points higher than 150 °C. The higher the total energy stored the best performance of the material at the same temperature (upper side of the chart) (Fig. 9).

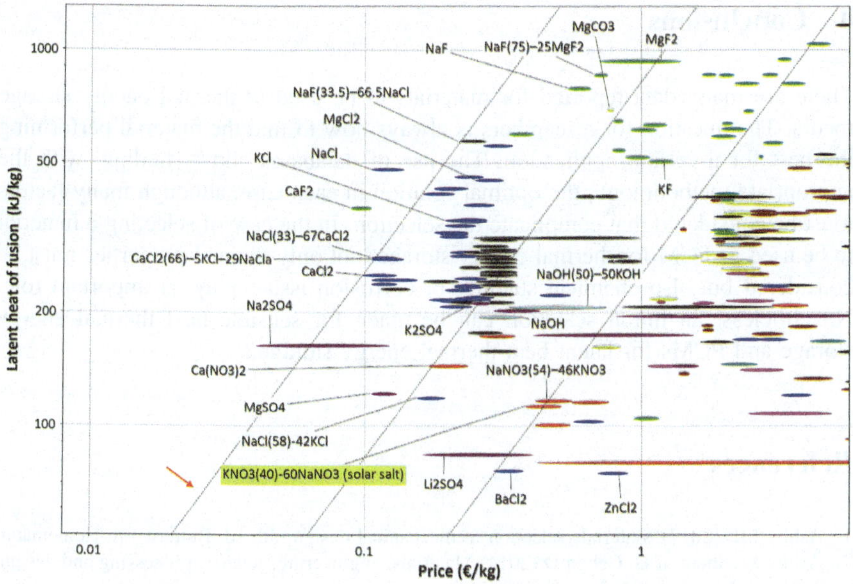

Fig. 8 Selection of PCMs for latent heat storage at temperatures higher than 150 °C. Chart from CES Selector 2018 [4]

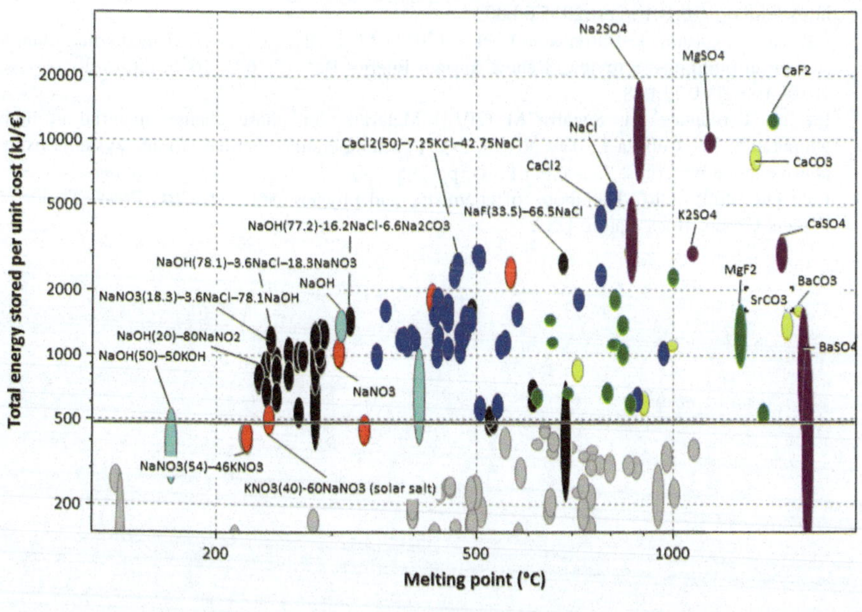

Fig. 9 Selection of PCMs for latent heat storage at temperatures higher than 150 °C. Chart from CES Selector 2018 [4]

4 Conclusions

There are many data reported for materials to be used as thermal energy storage media. The question for researchers is always how to find the material performing the best for a certain application. The use of databases allows finding, with the appropriate methodology, the optimal solution in each case; although many factors must be considered that complicate the selection. In the case of selecting a material to be used as PCM for thermal energy storage, not only thermal properties must be considered but also chemical stability or corrosion issues play an important role. Nevertheless, an initial selection can be made for sensible heat thermal energy storage and PCMs for latent heat thermal energy storage.

References

1. Ashby MF (2017) Materials selection in mechanical design, 5th ed. Butterworth-Heinemann
2. Ashby M, Shercliff H, Cebon D (2014) Materials: engineering, science, processing and design, 3rd edn. Butterworth-Heinemann, Oxford
3. Fernández AI, Martínez M, Segarra M et al (2010) Selection of materials with potential in sensible thermal energy storage. Sol Energy Mater Sol Cells 94:1723–1729. https://doi.org/10.1016/j.solmat.2010.05.035
4. Granta Design Ltd. (2018) CES Selector 2018
5. Barreneche C, Navarro ME, Cabeza LF, Fernández AI (2015) New database to select phase change materials: chemical nature, properties, and applications. J Energy Storage 3:18–24. https://doi.org/10.1016/j.est.2015.08.003
6. Cabeza LF, Castell A, Barreneche C et al (2011) Materials used as PCM in thermal energy storage in buildings: a review. Renew Sustain Energy Rev 15:1675–1695. https://doi.org/10.1016/j.rser.2010.11.018
7. Liu M, Fernández AI, Segarra M (2018) Materials for phase change material at high temperature. In: Cabeza L, Tay NHS (eds) High temperature thermal storage systems using phase change materials. Academic Press, pp 195–230
8. Lide DR (2005) CRC Handbook of chemistry and physics, 86th ed. CRC Press, Taylor & Francis Group, Boca Raton, USA

Part II
Experimental Characterization of Materials and Components for Thermal Energy Storage

Characterization of Materials for Sensible Thermal Energy Storage at High Temperature

Ángel G. Fernández, Laura Boquera and Luisa F. Cabeza

Abstract Among the several materials that can be appropriate for thermal energy storage (TES) application, this study focuses on sensible heat storage technology. In particular, are selected the common used materials at high temperature, molten salts and concrete. The aim is to define the key parameters and characterization techniques that determine the suitability of the material for TES. Broadly, in both materials thermo-physical and mechanical properties are evaluated. In the characterization of these materials highlights the procedure and equipment used to measure the specific properties. Mainly, the testing depends on the required sample size and operating temperature to be achieved. This chapter is based on the characterization carried out in previous researches as well as to define these thermal and mechanical techniques properly.

1 Introduction

One thermal energy storage technology is sensible heat storage, which takes advantage of heat capacity and the change in temperature of the material. The concept of sensible heat storage is that if temperature rises, energy is being absorbed and, the other way around, energy is released when temperature decreases [1]. Comparing sensible thermal energy storage (TES) to the other two technologies of heat storage (latent and chemical), it has a simpler design. Nevertheless, it also

Á.G. Fernández
Energy Development Center, University of Antofagasta,
Av. Universidad de Antofagasta 02800, Antofagasta, Chile

L. Boquera
CIRIAF-Interuniversity Research Centre on Pollution and Environment
Mauro Felli, Via G. Duranti 63, 06125 Perugia, Italy

L. Boquera · L. F. Cabeza (✉)
GREiA Research Group, INSPIRES Research Centre, Universitat de Lleida,
Pere de Cabrera s/n, 25001 Lleida, Spain
e-mail: lcabeza@diei.udl.cat

A. Frazzica and L. F. Cabeza (eds.), *Recent Advancements in Materials and Systems for Thermal Energy Storage*, Green Energy and Technology,
https://doi.org/10.1007/978-3-319-96640-3_6

has some disadvantages like the high volume of material required and that the temperature to store/release will not be constant. In sensible TES several storage mediums that could be either liquid or solid, such as water, oil, rock bed, bricks, sand or soil exist [2]. Since this TES technology has been considerably/noteworthy investigated in testing new materials, a methodology can be adapted to choose the optimum material [3–5]. Depending on the material properties a selection is made in relation to the desired application. A significant parameter that has influence on the decision of the storage medium is its operation temperature range. Previous studies were carried out about developing potential sensible TES materials for temperatures between 150 and 200 °C [2] and others between 500 and 750 °C [6]. Granta Design CES selector package allows a comparison of several materials, presented in bubble charts in function of different parameters. Operating/service temperatures above 500 °C are considered as high temperature conditions. In this chapter, sensible materials under high temperatures are assessed, focusing mainly on solar power plant applications. Broadly, regardless high temperature ranges/ exposure, to choose the proper material as a storage medium should fulfil some features. The requisites to be a material for TES application at high temperature from and economical and environmental point of view is to have abundant availability of resource as well as inexpensive/low-cost manufacturing of the material. Once a heat storage medium material with the previous mentioned characteristics is developed, the next step is to evaluate its thermo-physical and mechanical properties. The suitable thermo-physical properties to achieve are high values of thermal conductivity, heat capacity, density and specific heat capacity. Other two relevant properties are long-term thermal cycling and chemical stability, which are keys to determine the useful life of the system performance. Additionally, mechanical constraints such as good mechanical stability, low coefficient of thermal expansion, high fracture toughness and high compressive strength should be considered [6]. Moreover, the compatibility with heat transfer fluid and the heat exchanger design should be taken into account, ensuring good heat transfer between the heat transfer fluid and the storage material.

The amount of energy stored in the material is calculated the following Eq. 1.

$$Q = m \cdot c_{\mathrm{p}} \cdot \Delta T, \tag{1}$$

where

Q is the amount of heat stored in the material (J)
m is the mass of storage material (kg)
c_{p} is the specific heat capacity of the storage material (J/kg·K)
ΔT is the temperature change (K)

In this chapter focuses on two case studies. The characterization of molten salts and concrete are selected to study as sensible materials at high temperature. Molten salts are the most well-known material used for thermal energy storage in CSP plants while concrete has obtained promising results to have comparison.

2 Materials Used for Sensible Heat Storage at High Temperature. Molten Salts

2.1 Thermal Characterization Techniques

Sensible heat storage materials containing molten salts should be characterized in their operational range at high temperature in CSP plants. The main thermo-physical properties desired for TES materials in CSP are specific heat, melting/solidification enthalpy and temperature, thermal conductivity, thermal stability, density and viscosity. Each property can be experimentally obtained through the measurement techniques listed in Table 1.

Specific heat (C_p) is one of the key factors in the development of an efficiency storage system in a CSP plant, since this phenomenon controls the capacity of increase the temperature that can be stored or transported. The most widely used technique to measure C_p is the differential scanning calorimetry (usually done with a differential scanning calorimeter, DSC) and the ASTM 1269E is the most extended procedure to determine the specific heat of a material by DSC [7]. However, it should be also mentioned that new methodologies are arising to improve the accuracy and reduce the scattering of the results [8, 9]. In this direction, modulated DSC (MDSC) [1] is the most accurate technique for the measurement of C_p in molten salts in the range of temperatures between 200 and 400 °C [2]. MDSC measures the difference in heat flow between a sample and an inert reference as a function of time and temperature, however, a different heating profile (compared DSC conventional test) is applied to the sample and reference. Specifically,

Table 1 Thermo-physical characterization techniques of molten salts at high temperature

Measurement technique/equipment	Abbreviation	Temperature range	Parameter
Differential scanning calorimeter	DSC	−90 to 500 °C T > 500 °C needs special furnace and alumina crucibles	C_p, ΔH, Tmelting, T solidification
Modulated differential scanning calorimeter	MDSC	25 to 550 °C	C_p
Thermogravimetry	TGA	25 to 1200 °C	Degradation temperature
Differential thermal analysis	DTA	25 to 1200 °C	ΔT
Laser flash	–	−125 to 1100 °C	Thermal diffusivity
Reomether	–	−150 to 1000 °C	Viscosity
Pycnometer	–	25 to 500 °C	Density

a sinusoidal modulation (oscillation) is overlaid on the conventional linear heating or cooling ramp to yield a profile in which the average sample temperature continuously changes with time [3].

The heat capacity (C_p) of the sample is continuously determined by dividing the modulated heat flow amplitude by the modulated heating rate amplitude. This approach is based on the well-accepted procedures for determining C_p in conventional DSC. In conventional DSC, C_p is generally calculated (Eq. 2) from the difference in heat flow between a blank (empty pan) run and a sample run under identical conditions.

$$C_p = K \cdot C_p \cdot \frac{\text{Heat Flow(Sample)} - \text{Heat Flow(Blank)}}{\text{Heating rate}} \tag{2}$$

where $K \cdot C_p$ is the calibration constant

On the other hand, the heat capacity signal in modulated DSC is calculated based on a discrete Fourier transformation where the measured amplitudes of the sample temperature and heat flow modulation are compared to a reference sine wave of the same frequency. The equation used is

$$C_p = K \cdot C_p \frac{Q_{amp}}{T_{amp}} \frac{\text{Modulation period}}{2\pi} \tag{3}$$

where

C_p is the heat capacity
$K \cdot C_p$ is the heat capacity calibration constant
Q_{amp} is the heat flow amplitude
T_{amp} is the temperature amplitude

Another key thermal property of TES materials for CSP is the solidification/melting temperature. A high melting point results in more operation and maintenance costs in order to maintain the salt flowing through the pipes. The same technique that is used for specific heat (C_p) determination, the DSC, is used for obtaining the melting/solidification temperature.

Since in a CSP plant the higher operating temperature is related with a higher turbine efficiency, the property which plays a key role is the degradation temperature of the storage material. For a proper thermal stability study a thermo-gravimetric analysis (TGA) technique is commonly used.

As it can be seen in Fig. 1, thermal degradation tests of molten salts are usually carried out in a nitrogen atmosphere from room temperature to 600–700 °C using a heating rate about 10 °C/min. Depending on the molten salt composition, a dehydration process can be obtained (specially using calcium nitrate or a weight loss related with releasing NOx (usually observed between 220 and 250 °C).

Molten salt decomposition starts at 400 °C and the current criterion taken in order to fix the maximum thermal stability of the molten salts used like storage material in a CSP plant is contradictory in the literature. The most observed trend by

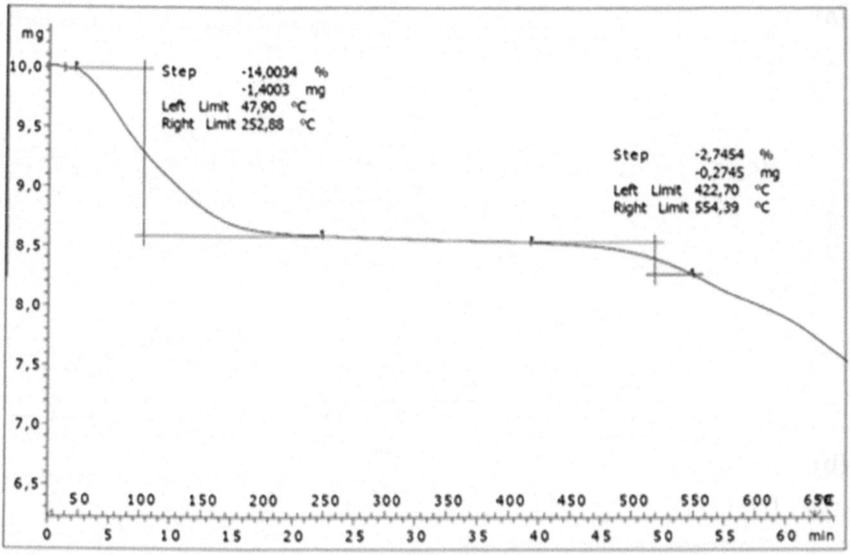

Fig. 1 Thermal decomposition curve of 48% $Ca(NO_3)2$ + 7% $NaNO_3$ + 45% KNO_3 mixture [10]

the authors working in this area [10–13] is to set the maximum operational temperature when molten salt has lost 3 mass%.

However, in a recent paper, Fernandez et al. [14] concluded that the maximum working temperature for the molten salts tested should be set at the onset of the TGA curve since salts continues losing weight once they have begun the decomposition process.

Table 2 shows this comparison in terms of decomposition temperature, comparing both criteria and the important difference between onset and 3 wt% temperature, about 100 °C.

This behaviour was confirmed through the isothermal test carried out in solar salt at 500 °C and at 450 °C (Table 2; Fig. 2).

Further research is needed in order to unify criterions since molten salt maximum operation temperature is one of the main important parameters for CSP design and operation.

Table 2 Mass loss in isothermal study performed at 450 °C

Molten salt	Initial mass (mg)	Mass lost (%)	Onset TGA temperature (°C)	3 wt% TGA temperature (°C)
60 wt% $NaNO_3$ + 40 wt% KNO_3	10.06	5.38	485.68	588.51
10 wt% $LiNO_3$ + 10 wt% Ca $(NO_3)_2$ + 60 wt% KNO_3 + 20 wt% $NaNO_3$	10.13	5.24	469.56	580.36

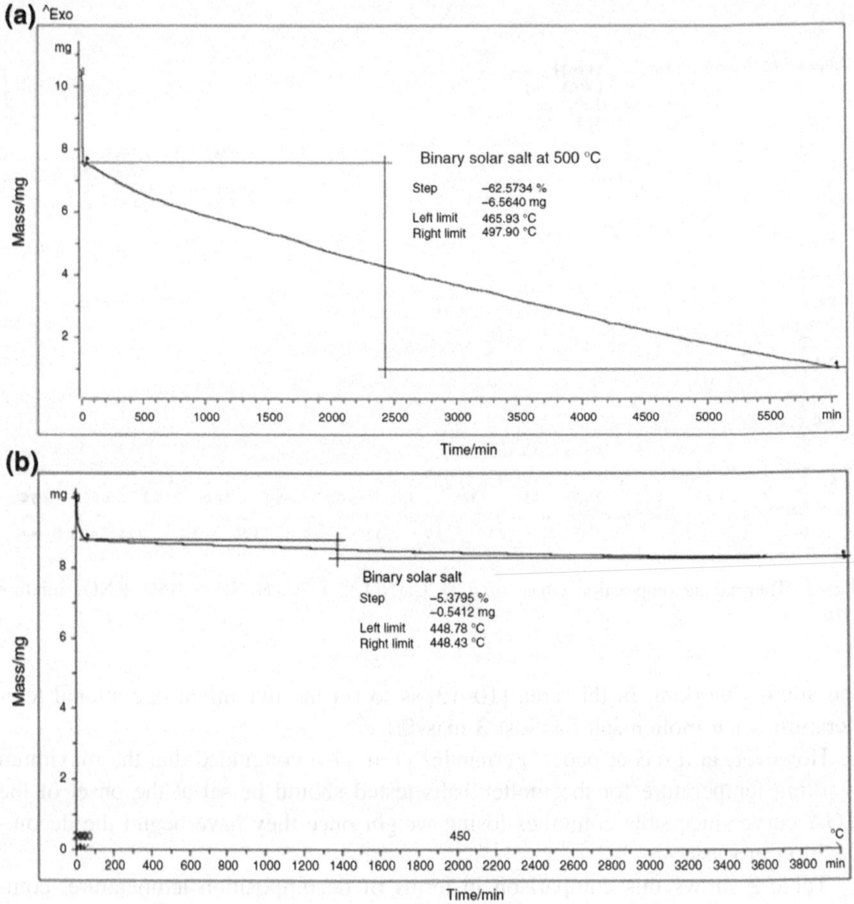

Fig. 2 Isothermal TGA test in solar salt. **a** At 500 °C, **b** 450 °C during 6000 min [14]

The technique differential thermal analysis (DTA) measures the temperature difference between a substance and a reference material as a function of the temperature during a controlled thermal cycle. It is a semi-qualitative technique that is usually used to perform the phase diagram in the molten salt mixture.

To obtaining the thermal conductivity in molten salts, usually the laser flash technique is used [15]. It is an indirect measurement since thermal diffusivity is the measured property, and knowing the density and specific heat, thermal conductivity can be calculated with Eq. 4:

$$k = a \cdot \rho \cdot C_{\mathrm{p}}, \tag{4}$$

where

k is the thermal conductivity
a is the thermal diffusivity
ρ is the density
C_p is the heat capacity

Some efforts are needed in order to develop new techniques for thermal conductivity measurement, since laser flash reduces the accuracy when molten salts are liquid, due to their volume variation at high temperature.

Viscosity (dynamic viscosity) of the molten salt needs to be known mainly because it affects the required energy consumption to pump it. A rheometer, usually a dynamic shear rheometer, is needed to measure the viscosity at certain temperature points and shear rates. Viscosity of molten salts is quite dependant to temperature and therefore enough measurements from above its melting/freezing point to the operating conditions are required. Note that rheological parameters, shear stress, deformation and shear rate, are calculated using conversion factors from the measured values, torque, deflection angle and speed, respectively. In a recent publication, Muñoz et al. [16] established that the rheological profile of molten salts is influenced by the rheometer geometry. The measurements carried out with the coaxial cylinder method showed that the samples have a Newtonian behaviour with a yield stress that needs to be exceeded to start the flow. On the other hand, when the measurements are carried out with parallel plate geometry, a shear thickening and a Newtonian behaviour are observed depending on the salt purity and the shear rate applied.

Density, similarly to thermal conductivity, is very sensitive to temperature. Therefore, it is highly recommended to measure it in the operating temperature range. A wide variety of methods are available for density measurement, i.e. hydrometers, pycnometers, digital density metres, hydrostatic balances. It is worth to point out that some density measurement methods determine apparent density (weight per volume) whereas others measure true density (mass per volume).

2.2 *Mechanical Characterization Techniques*

There are few studies regarding the mechanical tests performed in the salt properly. However, these kinds of tests are necessary for a better understanding in the recovery process from a freeze event in the molten salt during the operation.

Iverson et al. [17] carried out these mechanical tests focus on compressive and tensile strength, Young modulus and Poisson ratio.

Cylindrical samples are needed for the mechanical test following the ASTM Standard D4543 guidelines. In this case a block of salt is used (with a diameter around 5 cm) for the tensile strength method and a compressive stress is applied in one direction that induces tensile stress in an orthogonal direction. Iverson et al. performed these test at low temperature (solid solar salt) from 50 to 90 °C. It was found that as test temperature increase, compressive strength and Young´s modulus decrease for solar salt (60% $NaNO_3$ + 40% KNO_3).

As it was mentioned before, these mechanical tests performed directly in the solid salt are focus in special situations related with a freezing issue during the plant operation. On the other hand, mechanical problems using molten salt are mainly present in the container materials (carbon steel or stainless steels) at higher temperatures.

Mechanical tests typically use the yield strength as one of the main design parameters. However, it is well known that the mechanical properties of steels as the yield strength or the modulus of elasticity decrease with increasing temperature.

For a proper mechanical study in molten salt at high temperature, the slow strain rate technique is commonly used. The slow strain rate test (SSRT) involves a slow (compared to conventional tensile tests) dynamic strain applied at a constant extension rate in the environment of interest [18]. SSRTs are generally run until total failure, which may take hours or days but could occasionally be shorter, longer or interrupted. After failure, the specimens are examined for stress corrosion cracking (SCC), and measurements taken during and after the test are compared to similar data for an inert environment.

This type of tests is especially important in CSP industry since the thermal stress usually produced in the materials in contact with molten salts. Preuβner et al. [19] reported the following equipment for SSR test (Fig. 3).

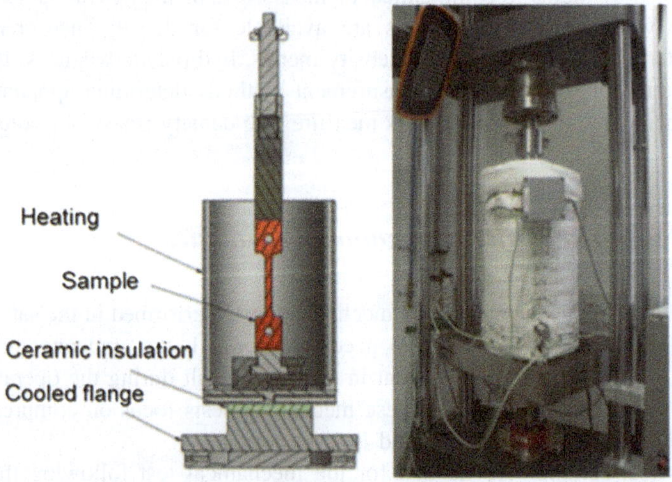

Fig. 3 Slow strain rate equipment for molten salts

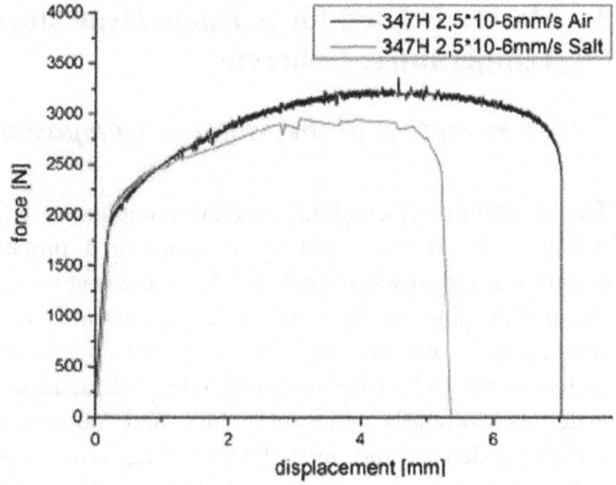

Fig. 4 Slow strain rate test performed in 347SS immersed in solar salt at 565 °C [21]

Slow strain rate tensile (SSRT) test is a standard testing method in which the specimen is subjected to a constant elongation rate [20]. During the test the material is exposed to an environment to investigate the resistance of metallic materials to environmentally assisted cracking (EAC) and its influence on the material fracture. Preuβner et al. [21] reported a reduction around 26% (Fig. 4) in the slow strain rate test performed in 347 stainless steel and carried out in air compare with solar salt (60 wt% $NaNO_3$ + 40 wt% KNO_3). Authors argue that microstructure changes during the test could be explained by nucleation and coarsening of carbides.

Cyclic mechanical loading with plastic deformation leads to micro-crack formation, and the presence of molten salts at high temperature can increase the crack growth and reduce the component life cycle.

The effect of salt exposure of alloys and their influence in the mechanical behaviour of components in CSP plants is one of the main important topics to solve in the next generation of CSP plants. Preuβner et al. [21] concluded that the exposure of materials to molten nitrate salts normally results in a loss of ductility and/or a reduction in strength. The alloys are usually prepared under a standard test procedure for mechanical testing after ASTM E8/E8M [22]. Mechanical properties (tensile strength, yield strength, elongation at break, elastic modulus) are usually determined through uniaxial tensile testing after molten salt mixture immersion. Following this standard, Prieto et al. [23] confirmed the reduction of yield strength and Young modulus as well as the elongation compared with testing at room temperature. A seven days creep-test experiment maintaining the sample (A516Gr.70) under working conditions (solar salt at 380 °C) and a load below the yield strength was performed. Changes in mechanical properties after the creep-test were confirmed, attributing the reduction of yield strength and elongation to the annealing of the simple.

3 Materials Used for Sensible Heat Storage at High Temperature. Concrete

3.1 *Description of the Material Composition and Behaviour*

The availability of essential concrete components is presented all over the world, making this ancient construction material a proper solid storage media. This compact stone material does not have environmental impact, it is workable and cheap. Focusing on the concrete composition, it is basically formed by cement, water, coarse and fine aggregates (gravel and sand). However, to meet specific requirements and to improve concrete performance "supplementary cementitious materials" (flay ash, silica fume, etc.) and "additives" (fluidificants) materials are usually added to the mix. Furthermore, other materials like polyethylene or polypropylene fibres may be included in order to solve some problems like permeability or to achieve a long-term stability [24].

To determine the concrete characterization for TES application, the study of concrete under high temperatures should be considered previously in order to understand its performance as temperature rises [25].

To show examples of sensible storage materials that can be appropriate as solid media, Khare et al. [6] highlight promising results in materials such alumina, geopolymer, high alumina concrete, and silicon carbide. The claimed energy stored values are 200, 298, 245 and 260 kJ/kg respectively, being concrete an interesting choice to deeply study.

A complete study of concrete as TES can be carried out at laboratory and later at pilot plant scale. This book chapter is focused on the characterization of the material at laboratory scale.

3.2 *Laboratory Characterization of Concrete for TES*

3.2.1 Experimental Procedure Considerations

To start an experimental study, the sample size and the replicates should be stated at the beginning. The considerations for material to test are established following standards. An example of European and American standards followed are EN 12350-1:2009 [26], EN 12390-1:2013 [27], EN 12390-2:2009 [28], ASTM C192/C192M [29]. Commonly, concrete mixtures are casted in plastic or metallic moulds. These moulds could be cubical either cylindrical. However, if particular flexural test is desired, prismatic beams/slabs are selected. Once the fresh concrete is casted in the mould, it is compacted with a vibrating table or pounding the mixture with a stick. These specimens remain in the mould during 24 h and after are demoulded. Succeeding, the samples are immersed in water, curing for 28 days in a conditioned room with temperature and relative humidity at 20 ± 2 °C and 90 ± 5%, respectively.

In the test procedure stands out the fact that most part of the properties should be analysed, before and after exposure to elevated temperatures. Above all, a decision factor that has influence in thermal energy storage is thermal stability to cycles. From studying the performance of concrete under thermal cycles the useful life/lifetime of the material is determined. Since, the purpose is to characterize the concrete to be the solid media of the storage tank in a CSP thermal power plant, the module should last as much as possible [24, 30, 25].

Regarding the specific constituents of concrete, Ozger et al. [31] had considered interesting to study aggregates and fibre properties. These properties could be known from the enterprise data sheet or whether by realizing some tests. This precise study of each component allows to have an idea of the materials behaviour separately from the whole mixture.

To characterize the concrete for TES applications, there are some principle parameters to analyse its performance under high temperatures. These properties to take into consideration are shown in Table 3.

3.2.2 Physical Characterization Techniques

The physical characterization of concrete is carried out first while the fresh concrete mixture is prepared and later when the concrete has hardened. The tests in fresh concrete are performed to determine the workability and setting time [32]. In thermal energy storage, the conventional analysis is the slump test in order to know the workability of the mixture. This previous parameter is usually checked following the European standard EN 12350-2:2009 [33] or the American standard ASTM C143 [34]. Some researchers as Adeyanju et al. [35, 36] and Ozger et al. [31] mentioned/considered the slump test results in their experiments.

Table 3 Properties to consider for concrete in TES application Adapted from Khare et al. [6]

Thermo-physical	Chemical	Mechanical	Environmental
Energy density Thermal conductivity Thermal diffusivity Specific heat capacity Heat capacity Density Porosity and vapour Permeability Thermal cycling Microstructure Mass loss Workability-consistency	Chemical changes Mineralogy Degradation	Compressive strength	CO_2 footprint
		Coefficient of thermal expansion	Energy consumption manufacture process
		Elastic modulus	
		Fracture toughness	

Once the concrete is hardened, a major factor that is considered from the beginning is doing a visual observation of the samples before and after exposure to elevated temperatures. This action shows the appearance of concrete, noticing if there is degradation in the sample surface (for instance, cracks or detachments of concrete mass). This procedure allows determining the evolution of fissures or colour changes with temperature increase. To deepen in concrete microstructure rather than doing a fist eye sight, scanning electron microscope (SEM) may be used. For the SEM analysis the standard EN ISO 17751-2:2016 [37] is used. Another basic characteristic to study is the density that can be determined following these standards EN 12390-7:2009 [38], ASTM C642–13 [39]. From the last standard mentioned, there is also two other interesting properties to measure, the porosity and the vapour permeability test according to the standard EN 15803:2009 [40]. From the literature, Girardi et al. [41] used Cembureau apparatus to determine gas permeability to air. Table 4 lists the equipment and characteristic required for the physical characterization techniques.

Concrete has a high density thanks the aggregates, which represent around 80% of the material mass. Hence, the aggregate selection will have a strong influence on the density. In usual concretes density values are near 2200–2400 kg/m^3, using calcareous or siliceous aggregates [41–44]. Figure 5 presents the mass loss of some concretes mixtures found in the literature which analyse how the density is reduced after thermal cycle's exposure to high temperatures. Figure. 5 shows that densities of concrete mixtures can decrease more or less depending on the dosage of components.

Table 4 Concrete physical characterization techniques at high temperature

Measurement technique/equipment	Parameter	Sample size	Standard	References
Scales	Density	–	EN 12390-7:2009 [38], ASTM C642 [39]	–
Slump test	Workability	Fresh concrete mixt		[31, 35, 36]
Vacuum, environmental chamber, cembureau apparatus	Porosity, vapour permeability, gas permeability to air	Cubic or cylindrical 40 × 40 mm 50 × 50 mm	EN 1936:2007 [42], EN 15803:2009 [40]	[41]
Scanning electron microscope (SEM) microscope	Fracture toughness	1–4 g	EN ISO 17751-2:2016 [37]	[31, 41]

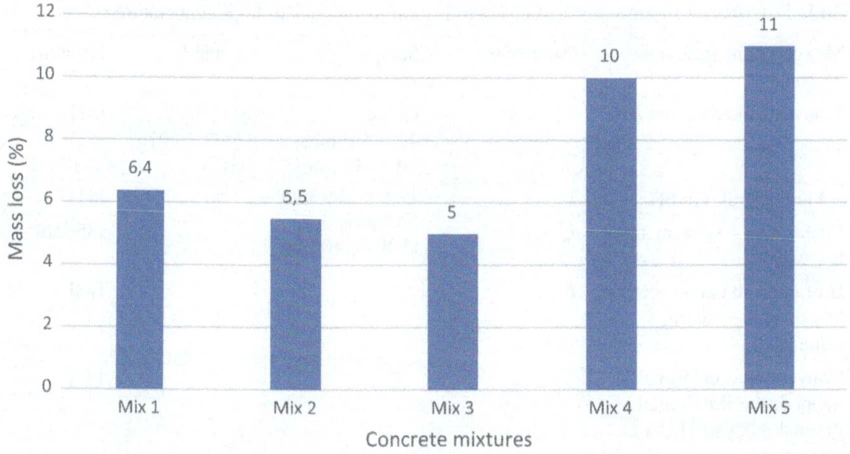

Fig. 5 Comparison of mass loss after exposure to 25–30 cycles up to 500–600 °C. Following are described the mixtures and their respective references. 1. Calcium aluminate cement (CAC) and blast furnace slag [43]. 2. Ordinary Portland cement (OPC) and polyethylene fibres (PE) [24]. 3. Ordinary Portland cement + polypropylene fibres (PP) + metallic fibres; mixture 26 [44]. 4. Ordinary Portland cement + Fly ash + polypropylene fibres (PP); mixture 15 [44]. 5. Calcium aluminate cement (CAC) + Fly ash + polypropylene fibres (PP); mixture 7 [44]

3.2.3 Thermo-Physical Characterization Techniques

As mentioned in the previous section, thermo-physical properties have a strong influence in the storage material. The essential properties are specific heat capacity, thermal conductivity, thermal diffusivity, and density. The equation that relates each other is

$$k = \alpha(T) \cdot c_{\mathrm{P}}(T) \cdot \rho(T) \tag{5}$$

A high thermal conductivity is desired to reduce the temperature gradient between charging and discharging phases, hence increasing the dynamic of the system [44]. The previous property along with higher heat capacity ($C_{\mathrm{p}} \cdot \rho$) and higher density reduce the volume of material required for a given storage capacity. Also, a high thermal diffusivity improves transient heat transfer and, hence, shortens the time for charging and discharging processes. These abovementioned properties values vary depending on temperature exposure.

To determine the thermal properties (thermal conductivity, thermal diffusivity and specific heat capacity), usually the transient plane source method is used. This is in accordance to ISO 22007-2:2016 [45]. A similar technique to measure thermal conductivity and thermal diffusivity was used by Adeyanju et al. [35, 36], named "Linear Heat Source method". Unlike the previous TPS method, this is destructive and it is required to drill the sample to introduce the needle to carry out the measurement. The standards that follow this test are UNE-EN ISO 8894-1:2010

Table 5 Concrete thermo-physical characterization techniques at high temperature

Measurement technique/ equipment	Parameter	Sample size	Standard	References
Transient method: hot disk	C_p, k, α	Cubic 40 × 40 mm 50 × 50 mm 100 × 100 mm Cylindrical Ø 40 × 40 mm	ISO 22007-2:2016 [45]	[31]
λ-Messtechnik GmbH	k		–	[31]
Linear heat source method: hot wire	k, α		–	[35, 36]
Half-open dynamic method based on the mixing principle	C_p		–	[48]
Conventional calorimetry method, the differential thermal analysis (DAT), and using the Law of mixtures			–	[44]
Differential scanning calorimeter (DSC)	C_p, ΔH, Tmelting, T solidification	10–40 mg	EN ISO 11357-1:2017 [49]	[41]
Thermo-gravimetric analysis (TGA)	Degradation temperature		ISO 11358-1:2014 [50]	[31]
	Mass loss		ASTM F495-99 (2011)	[31]
XRD- X-Ray	Mineralogy		EN 12698-2-2007 [51]	[41, 43]
Muffle furnace equipped with thermocouples (up to 1200 °C)	Thermal cycles resistance/ stability	Cubic 40 × 40 mm 50 × 50 mm 100 × 100 mm Cylindrical Ø 40 × 40 mm	–	[31, 41, 42, 43, 44]

[46] ASTM D5334-14 [47]. Table 5 summarizes the different equipment and parameters used to carry out the tests.

Laing et al. [24] measured the thermal conductivity using a steady-state comparative method using quartz glass as a reference material. In their research, a common concrete blending was used and limited thermal conductivities were obtained, from 1.45 W/m·K at ambient temperature to 1.20 W/m·K at 400 °C. On the other hand, Ozger et al. [31] analysed thermal conductivity with a k-metre EP500 by λ-Messtechnik GmbH at 25 °C, before and after an isothermal treatment. Also, in this research study they measured heat capacity using a DSC 92 Setaram differential scanning calorimeter over the range of 25–350 °C at a rate of 3 °C/min in N_2 using Al pans.

A major thermo-physical feature is thermal cycling resistance. This is key to ensure the lifetime of the material. John et al. [44] before heating thermal cycles, perform a conditioning entailed heating of the specimens at very low heating rates, 0.17 °C/min. up to 110 °C and subsequently at 0.67 °C/min up to 600 °C. After this previous treatment of the specimens, these are heated up to temperature for instance 200, 300, 400, 600, and 800 °C can be between 1 and 6 °C/min heating rate and exposed at desired temperature for 2 h in a muffle furnace. After the heating process, the specimens are removed from the furnace and allowed to cool naturally to the room temperature (20 ± 2 °C).

Regarding specific heat capacity, a part from the TPS test another technique was chosen by John et al. [44] using a simple calorimetry and also calculated using the Law of Mixtures. Whereas, Pan et al. [48] developed a half-open dynamic method based on the mixing principle. Also, Girardi et al. [41] using differential scanning calorimetry (DSC) following the standard EN ISO 11357-1:2017 [49].

Martins et al. [52] used thermo-gravimetric analysis (TGA) to measure the thermal stability of the "Heatcrete" material from EnergyNest company. After thermal cycles were performed up to 400 °C on the samples using a coupled to a Fourier transform infrared (FTIR) spectrometer [52, 53]. This TGA test follows the standard EN ISO 11358-1:2014 [50].

Usually the amount of concrete sample required for DSC, TGA, XRF and XRD is small (10–40 mg). Since concrete is a non-homogeneous material, these previous commented tests can be carried out also in concrete components. Tiskatine et al. [50] used a crushed sample of each aggregate type, sieved with an aperture of 100 mm to be transformed into a fine powder in order to measure the porosity and the specific heat and to perform the XRF, XRD and TGA/DTA analysis.

3.2.4 Mechanical Characterization Techniques

The properties analysed for mechanical resistance are basically compressive strength, flexural strength, elastic modulus and coefficient of thermal expansion. For these measurements prismatic samples (40 × 40 × 160 mm) are commonly used in order to carry out a flexural strength test in a first step, resulting in two cubic pieces following the standard EN 12390-5:2009 [54]. Afterwards, compressive strength is measured on the two cubes obtained from the prism sample according to the standard EN 12390-3:2009 [55]. Besides, the compressive test can be followed by the next standards ASTM C39/C39M [56], using cubic samples 15 × 15 cm or either cylindrical samples 10 × 20 cm. Mechanical parameters and techniques are shown in Table 6.

One of the main challenges for TES purpose are to evaluate and compare the mechanical outcome after thermal cycles upon high temperatures during a certain period of time. So far, the research studies focus on this parameter (known as "residual strength"), showing that the mechanical properties decrease with temperature [25].

Table 6 Concrete mechanical characterization techniques at high temperature

Measurement technique/ equipment	Parameter	Sample size	Standards	Literature
Universal testing machine	Compressive strength	Cubic 40 × 40 mm 50 × 50 mm 10 × 10 mm 15 × 15 mm Cylindrical Ø 40 × 40 mm Ø 100 × 200 mm	EN 12390-3:2009 [55]	[31]
	Flexural strength	40 × 40 × 160 mm	EN 12390-5 [54]	[31]
Universal testing machine + 3 transducers	Elastic modulus		ASTM C469 [57]	[41]
Dilatometer	Coefficient of thermal expansion (CTE)	Ø 100 × 200 mm	ISO 10545-8:2014 [61], EN 14581:2006 [62]	[31]

During thermal cycles, expansions and contractions are generated creating cracks in concrete. This explains the progressive decrease of ultra-pulse velocities (UPV) measured with heat cycles. This point is confirmed by the good correlation found between the compression strength and the UPV. Alonso et al. [41] estimated the dynamic elastic modulus of concrete from the UPV values, using Eq. 6

$$E = \frac{C_p^2 \rho (1+\nu)(1-2\nu)}{(1-\nu)}, \tag{6}$$

where E is the dynamic elastic modulus (GPa), C_p is the pulse propagation speed (m/s), ρ is the concrete density (kg/m^3), and ν is the Poisson coefficient.

Girardi et al. [41] used the universal testing machine and three displacement transducers to measure the vertical displacement to obtain the elastic modulus.

Figure 6 presents different concrete types and the results of compressive strength before and after exposure to high temperature. The mixture designs of these concretes are different, showing that a normal concrete with ordinary cement Portland and without supplementary cementiteous materials has the lower compressive strength. If some supplementary cementiteous materials are added or if the aggregate type is taken into account, the compressive strength increases.

Another property that its behaviour is also affected by the above-mentioned factors is the thermal expansion coefficient (CTE). This parameter characterizes the expansion of a material when heat is applied, defined as the expansion (shrinkage) of unit length of a material when the temperature of concrete is raised by one degree. When the material increase in length (elongate), the expansion is considered to be positive while shrinkage is produced is to be considered negative (shortens).

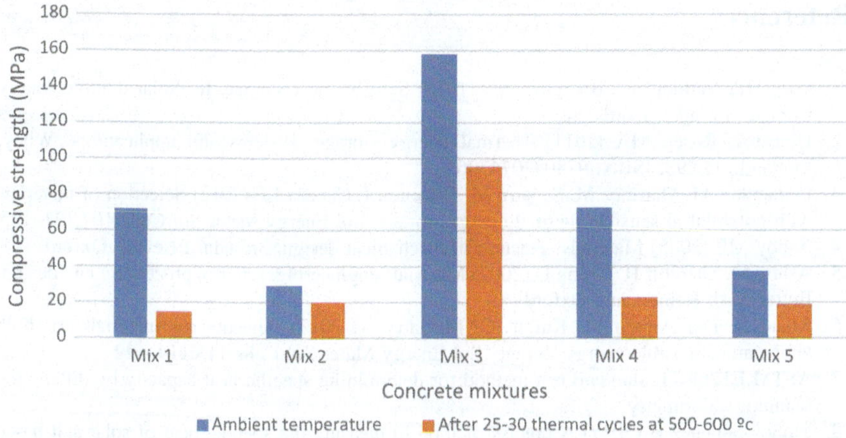

Fig. 6 Compressive strength and residual compressive strength of different concrete type. Following are described the mixtures and their respective references. 1. Calcium aluminate cement (CAC) and blast furnace slag [43]. 2. Ordinary Portland cement (OPC) and polyethylene fibres (PE) [24]. 3. Ordinary Portland cement + polypropylene fibres (PP) + metallic fibres; mixture 26 [44]. 4. Ordinary Portland cement + Fly ash + polypropylene fibres (PP); mixture 15 [44]. 5. Calcium aluminate cement (CAC) + Fly ash + polypropylene fibres (PP); mixture 7 [44]

Overall, the thermal expansion of concrete is studied through the dilatometric curve, which is a record of the fractional change of a linear dimension of a solid at a steadily increasing or decreasing temperature [60]. There are several standards that can be used to measure the CTE, such as: ISO 10545-8:2014 [61], EN 14581:2006 [62].

The most widely known techniques are by using a dilatometer. Ozger et al. [31] measured the thermal expansion through a Seiko Exstar TMA/SS 6000 dilatometer over the range of 20–550 °C at a rate of 3 °C/min in air with an alumina probe, repeating the measurement two more times. A horizontal dilatometer allows to monitor length changes of concrete during loading as a function of temperature and time. Temperature intervals can vary from 25 to 1000 °C, considering a heating rate 1 °C/min in a static air atmosphere.

The equation used to calculate the linear thermal expansion is the following:

$$\alpha = \frac{d(l/l_0)}{dT} = \frac{d\varepsilon}{dT}, \tag{7}$$

where α [1/°C] is the linear thermal expansion coefficient, l [m] the sample length at chosen temperature, l_0 [m] the initial sample length, ε [−] the thermal strain [63].

References

1. Garg HP, Mullick SC, Bhargava AK (1958) Sensible heat storage. In: Solar thermal energy storage. Springer, Dordrecht
2. Dinçer I, Rosen MA (2011) Thermal energy storage: systems and applications. Wiley, England. 93-298. ISBN: 9780470747063
3. Fernandez AI, Martinez M, Segarra M, Martorell I, Cabeza LF (2012) Selection of materials with potential in sensible thermal energy storage. Sol Energy Mater Sol Cells 107:131–135
4. Ashby MF (2005) Materials selection in mechanical design, 3rd edn. Elsevier, Oxford
5. Ashby M, Shercliff H, Cebon D (2007) Materials engineering, science, processing and design. Butterworth-Heinemann, Oxford
6. Khare S, Dell'Amico M, Knight C, McGarry S (2013) Selection of materials for high temperature sensible energy storage. Sol Energy Mater Sol Cells 115:114–122
7. ASTM E1269-11. standard test method for determining specific heat capacity by differential scanning calorimetry
8. Muñoz-Sánchez B (2017) A precise method to measure the specific heat of solar salt-based nanodluids. J Therm Anal Calorim 129:905–914
9. Ferrer G, Barreneche C, Solé A, Martorell I, Cabeza LF (2017) New proposed methodology for specific heat capacity determination of materials for thermal energy storage (TES) by DSC. J Energy Storage 11:1–6
10. Fernandez AG, Ushak S, Galleguillos H, Perez F (2014) Development of new molten salts with $LiNO_3$ and $Ca(NO_3)_2$ for energy storage in CSP plants. Appl Energy 2014(119):131–140
11. Ren N, Wu Y-T, Ma C-F, Sang L (2014) Preparation and thermal properties of quaternary mixed nitrate with low melting point. Sol Energy Mater Sol Cells 127:6–13
12. Federsel K, Wortmann J, Ladenberger M (2015) High-temperature and corrosion behavior of nitrate nitrite molten salt mixtures regarding their application in concentrating solar power plants. Energy Procedia 69:618–625
13. Gimenez P, Fereres S (2015) Effect of heating rates and composition on the thermal decomposition of nitrate based molten salts. Energy Procedia 69:654–662
14. Fernandez AG, Veliz S, Fuentealba E, Galleguillos H (2017) Thermal characterization of solar salts from north of Chile and variations of their properties over time at high temperature. J Therm Anal Calorim 128:1241–1249
15. Smirnov MV, Khokhlov VA, Filatov ES (1987) Thermal conductivity of molten alkali halides and their mixtures. Electrochim Acta 32(7):1019–1026
16. Muñoz-Sánchez B, Nieto-Maestre J, Veca E, Liberatore R, Sau S, Navarro H, Ding Y, Navarrete N, Juliá JE, Fernández AG, García-Romero A (2018) Rheology of solar-salt based nanofluids for concentrated solar power. Influence of the salt purity, nanoparticle concentration, temperature and rheometer geometry. Sol Energy Mater Sol Cells 176:357–373
17. Iverson BD, Broome ST, Siegel NP, Temperature dependent mechanical property testing of nitrate thermal storage salts. Sandia Report 2010-5481C
18. Henthorne M (2016) The slow strain rate stress corrosion cracking test—a 50 year retrospective. Corrosion 72:1488–1518
19. Preußner J, Pfeiffer W, Piedra E, Oeser S, Tandler M, von Hartrott P, Maier G (2017) Mechanical tests in liquid molten salts. RAISELIFE Dissemination Workshop, Madrid, May 17, 2017
20. ASTM G129-00 (2006)
21. Preußner J, Peiffer W, Piedra E, Oeser S, Tandler M, von Hartrot P, Maier G (2016) Long-term material tests in liquid molten salts. Advances in materials technology for fossil power plants. In: Proceedings from the eighth international conference, Albufeira, Algarve, Portugal, 11–14 Oct 2016
22. ASTM A370-16 (2016) Standard test methods and definitions for mechanical testing of steel products

23. Prieto C, Barreneche C, Martínez M, Cabeza LF, Fernández AI (2018) Thermomechanical testing under operating conditions of A516Gr70 used for CSP storage tanks. Sol Energy Mater Sol Cells 174:509–514
24. Laing D, Bahl C, Bauer T, Fiss M, Breidenbach N, Hempel M (2012) High-temperature solid-media thermal energy storage for solar thermal power plants. Proc IEEE 100:516–524
25. Laing D, Steinmann W-D, Tamme R, Richter C (2006) Solid media thermal storage for parabolic trough power plants. Sol Energy 80:1283–1289
26. John EE, Hale WM, Selvam RP (2010) Effect of high temperatures and heating rates on high strength concrete for use as thermal energy storage. In: ASME 2010 4th International Conference on Energy Sustainability, Volume 2 Phoenix, Arizona, USA, May 17–22, 2010
27. EN 12350-1:2009 Testing fresh concrete—Part 1: Sampling
28. EN 12390-1:2013 Testing hardened concrete—Part 1: Shape, dimensions and other requirements for specimens and moulds
29. EN 12390-2:2009. Testing hardened concrete—Part 2: Making and curing specimens for strength tests
30. ASTM C192/C192M, Standard practice for making and curing concrete test specimens in the laboratory
31. Ozger OB, Girardi F, Giannuzzi GM, Salomoni VA, Majorana CE, Fambri L, Baldassino N, Di Maggio R (2013) Effect of nylon fibres on mechanical and thermal properties of hardened concrete for energy storage systems. Mater Des 51:989–997
32. ASTM C403. Test Method for Time of Setting of Concrete Mixtures by Penetration Resistance
33. EN 12350-2:2009. Testing fresh concrete. Slump-test
34. ASTM C143. Standard Test Method for Slump of Hydraulic-Cement Concrete
35. Adeyanju AA, Manohar K (2011) Effects of steel fibres on properties of concrete for energy storage application. J Eng Appl Sci 6:336–345
36. Adeyanju AA, Manohar K (2011) Effects of steel fibers and iron filings on thermal and mechanical properties of concrete for energy storage application. J Miner Mater Charact Eng 10:1429–1448
37. EN ISO 17751-2:2016. Scanning electron microscope method
38. EN 12390-7:2009 Testing hardened concrete—Part 7: Density of hardened concrete
39. ASTM C642-13. Standard test method for density, absorption, and voids in hardened concrete
40. EN 15803:2009. Conservation of cultural property—test methods—determination of water vapour permeability (dp)
41. Girardi F, Giannuzzi GM, Mazzei D, Salomoni V, Majorana C, Di Maggio R (2017) Recycled additions for improving the thermal conductivity of concrete in preparing energy storage systems. Constr Build Mater 135:565–579
42. EN 1936:2007. Natural stone test methods—Determination of real density and apparent density, and of total and open porosity
43. Alonso MC, Vera-Agullo J, Guerreiro L, Flor-Laguna V, Sanchez M, Collares-Pereira M (2016) Calcium aluminate based cement for concrete to be used as thermal energy storage in solar thermal electricity plants. Cem Concr Res 82:74–86
44. John E, Hale M, Selvam P (2013) Concrete as a thermal energy storage medium for thermocline solar energy storage systems. Sol Energy 96:194–204
45. Gil A, Medrano M, Martorell I (2009) State of the art on high temperature thermal energy storage for power generation. Part 1: Concepts, materials and modellization. Renew Sustain Energy Rev 14:31–55
46. ISO 22007-2:2016. Plastics—determination of thermal conductivity and thermal diffusivity—Part 2: Transient plane heat source (hot disc) method
47. EN ISO 8894-1:2010. Refractory materials—determination of thermal conductivity—Part 1: Hot-wire methods (cross-array and resistance thermometer)
48. ASTM D5334-14. Standard Test Method for Determination of Thermal Conductivity of Soil and Soft Rock by Thermal Needle Probe Procedure

49. Pan J, Zou R, Jin F (2017) Experimental study on specific heat of concrete at high temperatures and its influence on thermal energy storage. Energies 10:33
50. EN ISO 11357-1:2017. Differential scanning calorimetry (DSC) methods for the thermal analysis of polymers. General principles
51. Martins M, Villalobos U, Delclos T, Armstrong P, Bergan PG, Calvet N (2015) New concentrating solar power facility for testing high temperature concrete thermal energy storage. Energy Procedia 75:2144–2149
52. Hoivik N, Greiner C, Bellido E, Barragan J, Bergan P, Skeie G, Blanco P, Calvet N (2017) Demonstration of EnergyNest thermal energy storage (TES) technology. AIP Conf Proc 1850:080011
53. Tiskatine R, Aharoune A, Bouirden L, Ihlal A (2017) Identification of suitable storage materials for solar thermal power plant using selection methodology. Appl Therm Eng 117:591–608. EN ISO 11358-1:2014. Plastics—thermogravimetry (TG) of polymers—Part 1: General principles
54. EN 12698-2-2007. Chemical analysis of nitride bonded silicon carbide refractories—Part 2: XRD methods
55. EN 12390-5:2009. Testing hardened concrete—Part 5: Flexural strength of test specimens
56. EN 12390-3:2009. Testing hardened concrete—Part 3: Compressive strength of test specimens
57. ASTM C39/C39M. Standard Test Method for Compressive Strength of Cylindrical Concrete Specimens
58. ASTM C469. Standard Test Method for Static Modulus of Elasticity and Poisson's Ratio of Concrete in Compression
59. ASTM F495-99. Standard Test Method for Weight Loss of Gasket Materials upon Exposure to Elevated Temperatures
60. Kodur V, Properties of concrete at elevated temperatures. Hindawi Publishing Corporation, ISRN Civil Engineering, vol 2014, Article ID 468510, 15 pages
61. ISO 10545-8:2014: Ceramic tiles—Part 8: Determination of linear thermal expansion
62. EN 14581:2006: Natural stone test methods—Determination of linear thermal expansion coefficient
63. Pavlík Z, Fořt J, Pavlíková M, Pokorný J, Trník A, Studnička J, Čítek D, Kolísko J, Černý R (2016) High temperature exposure of HPC—experimental analysis of residual properties and thermal response. MATEC Conf 63:01004

Experimental Methods for the Characterization of Materials for Latent Thermal Energy Storage

Camila Barreneche, Anna Laura Pisello, Ana Inés Fernández and Luisa F. Cabeza

Abstract This chapter summarizes the techniques to characterize phase change materials for thermal energy storage applications. The most relevant properties include giving answer to physical, thermal, and technical requirements. The chapter includes characterisation at different scales, from laboratory (including latent heat, melting temperature, specific heat, thermal conductivity, durability and cyclability) to pilot plant scale, including a novel dynamic characterization of composites.

1 PCM Most Relevant Properties

Phase change materials (PCM) characterization is one of the key steps when designing a PCM storage system because the compatibility between materials, the reliability, and the performance under working conditions must be assured during the all life of the final system.

C. Barreneche · A. I. Fernández
Department of Materials Science and Physical Chemistry, Universitat de Barcelona, Martí i Franqués 1, 08028 Barcelona, Spain

C. Barreneche
School of Chemical Engineering, BCES (Birmingham Centre of Energy Storage), University of Birmingham, Birmingham B15 2TT, UK

A. L. Pisello
CIRIAF—Interuniversity Research Centre on Pollution and Environment "Mauro Felli", Via Duranti 63, 06125 Perugia, Italy

A. L. Pisello
Department of Engineering, University of Perugia, Via Duranti 93, 06125 Perugia, Italy

L. F. Cabeza (✉)
GREiA Research Group, INSPIRES Research Centre, Universitat de Lleida, Pere de Cabrera s/n, 25001 Lleida, Spain
e-mail: lcabeza@diei.udl.cat

A. Frazzica and L. F. Cabeza (eds.), *Recent Advancements in Materials and Systems for Thermal Energy Storage*, Green Energy and Technology,
https://doi.org/10.1007/978-3-319-96640-3_7

Table 1 PCM requirements to be used in a real application

	Physical and technical requirements	Thermal requirements
LHTES through PCM	– Low density variation and small volume change – High energy density – Small or no subcooling – No phase segregation – Low vapour temperature – Chemical and physical stability – Compatible with other materials	– Suitable phase change temperature fitted to application – Large phase change enthalpy (ΔH) and specific heat (C_p) – High thermal conductivity (except for passive cooling) – Reproducible phase change – Thermal stability – Cycling stability

For that reason, it is important to remark the main properties required which have been mentioned by Cabeza et al. [1] and these properties are listed in the next Table 1.

Once the requirements for materials to be used for thermal energy storage are defined, it is possible that a material cannot accomplish all these properties. While all the mentioned properties are important, the thermal requirements are the most significant.

There are several criteria that limited the storage process from the beginning and are listed as follows:

- The first one is the segregation and it is defined as the separation between phases when one material is not homogeneous. If the PCM is separated in two parts, its thermophysical properties will change drastically. Therefore, this effect must be studied and completely avoid before the implementation.
- The second one is the subcooling. One substance presents subcooling when it exists at liquid state even at those temperature levels while it is supposed to be solid. Then, it overpasses the melting temperature being liquid, and this is one way to detect the subcooling.
- Corrosion problems between PCM and containers are the third main issue to be considered.
- Physicochemical stability at long term: one substance used as PCM must be stable over cycles in order to retain their thermophysical properties allowing the well-operation of the final system.
- The leakage is one of the most important drawbacks because if this effect appears, the effectivity of the system will decrease drastically, thus it have to be studied before the PCM implementation.

A distinction is needed among the material properties that can be studied and characterized in the laboratory either from bulk PCM samples and encapsulated PCM samples, from those properties that should be characterized in the PCM composite for a passive system or as part of a PCM slurry in an active system. That means that sample size is relevant and in these cases goes from few milligrams to several grams. The performance of the PCM in these systems results in the combination of several properties and characteristics not only of the PCM itself, but from the whole system that requires experimental characterization at bigger scale.

2 Characterization of PCM in the Laboratory

Thermophysical properties of PCM are the main design criteria for PCM systems. Among them the latent heat or melting-solidification enthalpies ΔH_m, the melting temperature Tm and the temperature range of the phase change, the specific heat capacity, C_p, and the thermal conductivity (k) are the most relevant.

2.1 Latent Heat

As mention before, the latent heat absorbed/released during the phase change is one of the key properties to be studied and considered when implementing a phase change material in such an application.

The most used and most powerful technique to measure the phase change enthalpy or latent heat absorbed/released by a phase change material is the differential scanning calorimetry [2]. A DSC is a thermo-analytical technique in which the difference in the amount of heat required to increase the temperature of a sample and a reference is measured as a function of temperature. There are several criteria to be considered when characterizing the phase change enthalpy of a PCM:

- Type mode: there are several experimental modes to programme a DSC. Barreneche et al. [3] studied and compare the most used DSC modes to characterize PCM. The conclusion was that for all common PCM the most used must be the dynamic mode which consists of applying a constant heating rate from the initial to the final temperature.
- Heating rate: the lower the heating rate, the better the DSC signal because most thermal equilibrium states will be reached during the DSC mode as explained by Melhing et al. [4].
- Temperature range must be selected taking into account the phase change temperature and must be around the half-measurement. Therefore, the best programmed temperature method to measure the phase change enthalpy is from 10 °C before the phase change up to 10 °C after the phase change [5].
- The sample mass selected to analyse a phase change material must be constant over the samples under study since the signal will be mass-dependent as was studied by Gunter et al. [6].

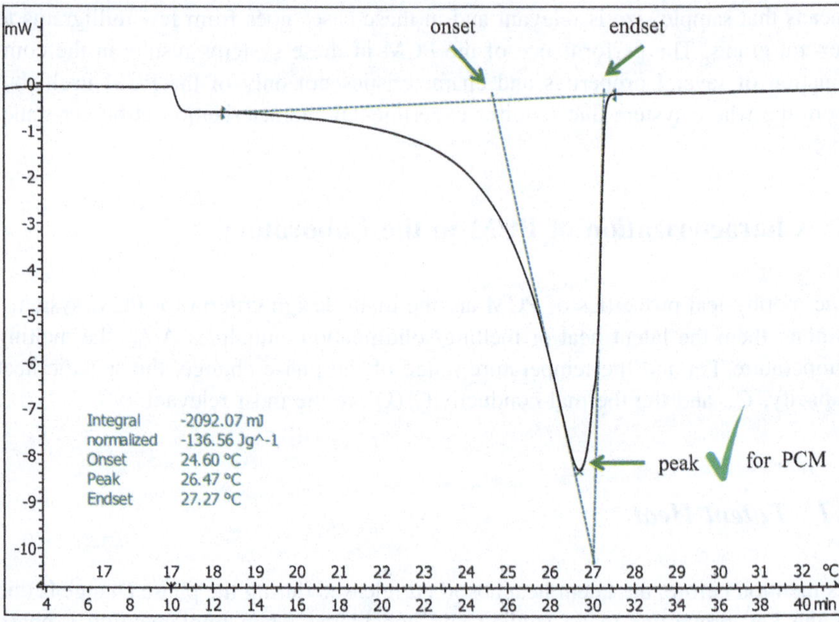

Fig. 1 Type of defined temperature for phase change material characterization

2.2 *Melting Temperature*

There are several methods to characterize the melting/solidification temperature of a PCM, but again the DSC is the most used to obtain proper results. Therefore, there are four temperatures defined as important parameters and are described in Fig. 1:

- Peak temperature: it is the maximum of the energy-process peak.
- On-set temperature: it is the starting temperature of the energy-process.
- End-set temperature: it is the finishing temperature of the energy-process.
- Temperature range: gradient temperature between the starting and ending temperatures energy-process (from on-set to end-set).

2.3 *Specific Heat—C_p*

The C_p is also important to be measure for the PCM characterization, especially in solid and liquid states (before and after the phase change transition).

There are several methods to measure the C_p and Ferrer et al. [7] compared them and state several conclusions. The three methods compared are the following which are shown in Fig. 2:

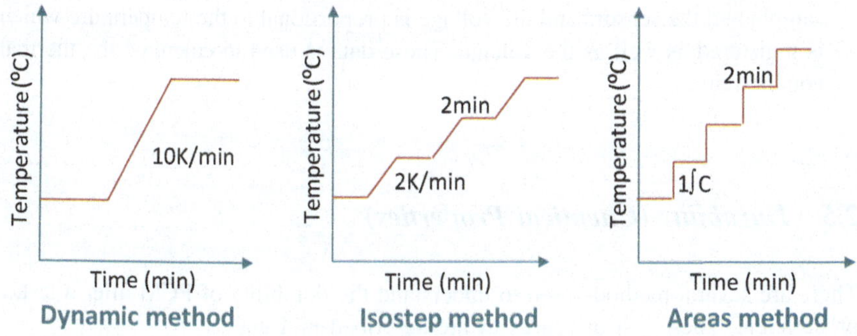

Fig. 2 DSC methods to be used to C_p analyses

- Dynamic method: a temperature range is established to measure the C_p. A heating rate is applied as fast as possible considering the lowest amount of mass as possible to be analysed.
- Isostep method: the temperature is increased 1 °C by 1 °C by 2 min isothermal steps and a heating rate of 2 K/min is applied between these isothermal steps.
- Areas method: the temperature is increased 1 °C by 1 °C without a heating rate between them. It is the simplest method but is the one with the best performance as state by Ferrer et al. [7].

2.4 Thermal Conductivity

The thermal conductivity property is directly related with the heat transfer during the charge/discharge process of the PCM. There are several methods to measure this PCM property but there are three methods which are the main used and applied:

- Laser flash [8]: this method is able to measure conductive materials; the ones with low thermal conductivity must be measure with other available techniques. This technique consists of applying a laser pulse and measure the heat output through the sample thickness by also registering the temperature gradient. It is mostly used for solid materials but some authors apply this technique to measure liquid ones.
- Hot-wire [9]: this method must be applied only for electric insulator materials and it is able to measure the thermal conductivity from 10 up to 150 °C. This technique consists on applying a temperature gradient to the sample and registers the thermal output that this heat produces. This technique is able to measure liquids, bulk solids, and granular solids.
- Hot disk [10]: this technique is able to measure liquids, solids and powder materials. It is also known as Gustafsson probe. It consists of a concentric ring where is located the heater. Thereby, a constant pulse is applied between the

sample and the sensors and the voltage is proportional to the temperature which is registered as well as the voltage. These data is used to calculate the thermal conductivity.

2.5 Durability (Chemical Properties)

There are several methods used to understand the durability of PCM after a test, a process, etc. These test or processes are described in Table 2.

Durability is defined [11] as a key attribute centred for security and economy aspects of the products. It is difficult to characterize, quantify and use for the material selection because it depends on the service conditions and there are several mechanisms that act under general or particular conditions due to material type and ambient conditions. Moreover, the material combination and configuration will be restrictive in the material role to play. Therefore, the durability of a material is the capacity of a material to conserve the characteristics and functionality for such application during all planned service life [12].

The techniques used to follow and control the durability of a material are as follows:

- FT-IR: infrared spectroscopy is one of the main used and most versatile techniques used to follow the chemical degradation after a process/test. This technique considers the vibration of each chemical bound when an infrared light is applied. The obtained signal is a characteristic peak which is able to be identified as one chemical substance (or a family of substances). If one characteristic peak disappears or appears is a crucial signal that the material/substance is under a degradation process [13]. This technique is applied basically for organic materials.
- XRD: X-ray diffraction is another power technique to follow chemical degradation when a material is under a process/test. This technique consists of diffract the x-ray when these x-ray are applied to a substance. Each signal obtained is

Table 2 Materials test which can produce PCM degradation

TEST	Definition
Ageing test	To apply a temperature, light, or any external agent that has the ability to degrade the material after a period of time under this ambient/artificial factor
Corrosion test and compatibility test	To apply a corrosive ambient or degradation environment to the PCM in order to understand if this material is degraded by corrosion effect
Flammability test	To burn a PCM under an extra-oxidation conditions in order to follow the flammability of the material as well as the material characteristics of the before and after the burning process

characteristic for each substance/material and if a degraded process is occurring the diffractogram will drastically change. Barreneche et al. [13] used XRD technique to follow PCM degradation. This technique can be used for inorganic and organic materials.
- Chemiluminiscence results from the conversion of chemical energy into light which takes place in many elementary reactions of more complex reaction mechanisms [14]. This technique has been used to follow the chemical degradation during oxidation the molecules of a reaction product. It is basically used for organic materials.

2.6 Cyclability

The cyclability of a PCM is defined as the stability of PCMs after several thermal or mechanical cycles.

Thermal cycling test consists on applying a cycling temperature range passing through the phase change temperature and followed by characterizing the thermal and the chemical properties before and after the thermal cycles [15].

Mechanical cycling test makes sense in active systems when a PCM slurry is pumped through a circuit and the thermal and mechanical properties are measure before and after the mechanical cycles [16].

Thermal cycling stability is followed and controlled thermally by measuring the phase change enthalpy with similar methods than the ones explained in Sect. 2.1 of this Chapter.

Chemical cycling stability is followed and controlled chemically by analysing the composition of the PCM before and after thermal test with the techniques described in Sect. 2.5 of this chapter.

3 Novel Dynamic Characterization of Composites for Energy Engineering Applications

By further enlarging the prospective of PCM applications, and considering engineering scale ones such as the ones in the built environment or other energy systems and technologies, the characterization of PCM composite represents a particular scenario since the amount of PCM within the composite material will be relatively low (i.e. around 10 wt%) compared to the pure-material analysis. The techniques for characterization to be used are almost the same than explained in sections above but, there are several novel dedicated methodologies own-developed in order to characterize them [14]. Moreover, specific methodologies have been therefore developed to characterize composite materials starting from the knowledge developed at pure-material level, such as the one described for example, by Barreneche et al. [17].

The characterization of TES materials for buildings or other applications becomes a crucial issue even when those materials are coupled or integrated to other materials or systems, which may be responsible for PCM variability in performance. These systems may be for instance a concrete with PCM inside, a floor slab to be installed into an envelope complex stratigraphy, a storage tank combining several fluids and container materials themselves. All these solutions need a further dynamic characterization which most typically is carried out by means of numerical methods trying to simulate the TES effect integrated into their realistic boundaries, and the other materials effect at the same time. Given the complex issue to be faced, those simulation models are not always correctly validated as a whole, but they use validated data for each material simulation. Typically, as previously stated, each material analysis is carried out at lab scale in very small sample quantities and sizes, which may be not enough representative of a real composite material for engineering application. And, at the same time, the basic characterization still is not enough technologically advanced to be carried out in a real pilot plant or into a real case study building. In this perspective, a new scale characterization framework is going to be properly defined with the threefold purpose to

- investigate real dynamic conditions of small engineering scale composite materials and systems, by also
- reducing experimental time and costs, and
- better controlling boundary conditions compared to field experiments.

This sort of intermediate scale characterization is therefore much needed in case of dynamically variable materials where environmental fluctuations play a key role in determining their behaviour in the field, and where an optimal control of such properties is needed in combination to other materials or systems, i.e. composites or multidimensional technologies.

Since there are no international standards regulating such methods, they are mainly focused on ad-hoc procedures developed by scientists and implemented at industrial level. Therefore, this technical policy gap has to be highlighted in this field, compared to the chemical small-scale characterization and the large-scale industrial characterization levels described in this chapter.

The novel characterization method basically consists of the integration of an environmental simulation chamber simulating realistic dynamic profiles of air temperature, air velocity, relative humidity, solar radiation to a complex temperature measurement system equipped in all the key points of the sample, and eventually a hot disk apparatus reproducing the transient plane source technique [18–20].

This combined innovative method has been applied for the first time to investigate classic thermal insulation materials for building envelopes, where the effect of ambient temperature fluctuation is investigated in dynamic realistic conditions re-created within the experimental apparatus. The purpose was to quantify the variability of thermal performance of such classic materials, which are typically considered as un-variable materials for building applications, with constant thermal conductivity, diffusivity and specific heat. The selected materials corresponded to expanded polystyrene (EPS), extruded polystyrene (XPS) and expanded polyurethane (PUR), specifically selected because of their (i) construction typicality, (ii) closed pore structure, that minimizes their hygro-thermal fluctuation in real conditions. The experiment demonstrated that, despite these premises, also these materials present key variation of their performances in realistic environmental conditions, demonstrating how the combined experimental method may be much needed while investigating building or energy system dynamics, by simulating time-varying boundary operations. The sample positioning within the chamber is reported in Fig. 3 (upper part), while the simulated cycles are reported in Fig. 3 (lower part).

The experiment allowed identifying very consistent thermal variability patterns (Fig. 4, left) while slowly increasing and decreasing ambient temperature in the test. The whole variability showed non-negligible difference in thermal parameters with varying the environmental dynamics (Fig. 4, right) confirming the necessity to reproduce realistic conditions in the lab, for reliable quantitative thermal evaluation.

Even more interesting results were achieved while investigating the dynamic performance of TES-doped composite materials. Also this experiment dealt with building applications, since it was aimed at studying the dynamic behaviour of structural concrete for building envelope component, but replicability is ensured by simply varying the test environmental conditions where reproducing the working boundaries of each tested material. Given the intrinsically adaptable behaviour of thermal energy storage systems such as phase change materials, these experimental characterization methods allowed to identify the true transition phase, duration, absorbed/released energy of the whole composite material at engineering scale. Figure 5 (left) reports the sample positioning within the environmental simulation chamber that this time is only equipped with a huge number of thermal probes for

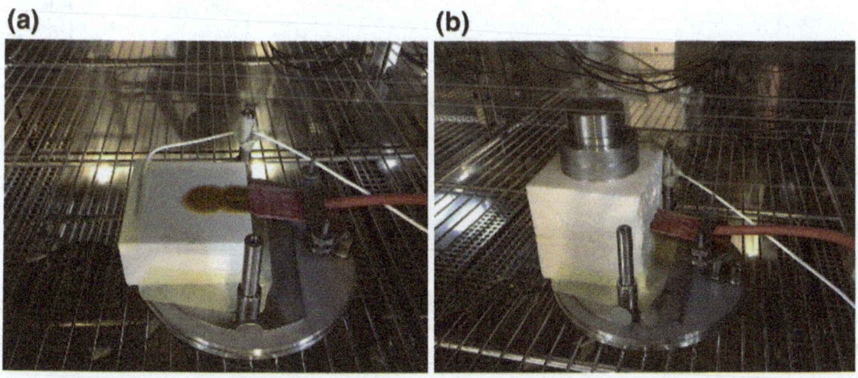

Fig. 3 Positioning of the tested material within the chamber and equipped with the Transient Plane Source Technique (hot disk) sensor (upper part). Dynamic thermal fluctuation cycles simulated within the experimental apparatus (bottom part) [21]

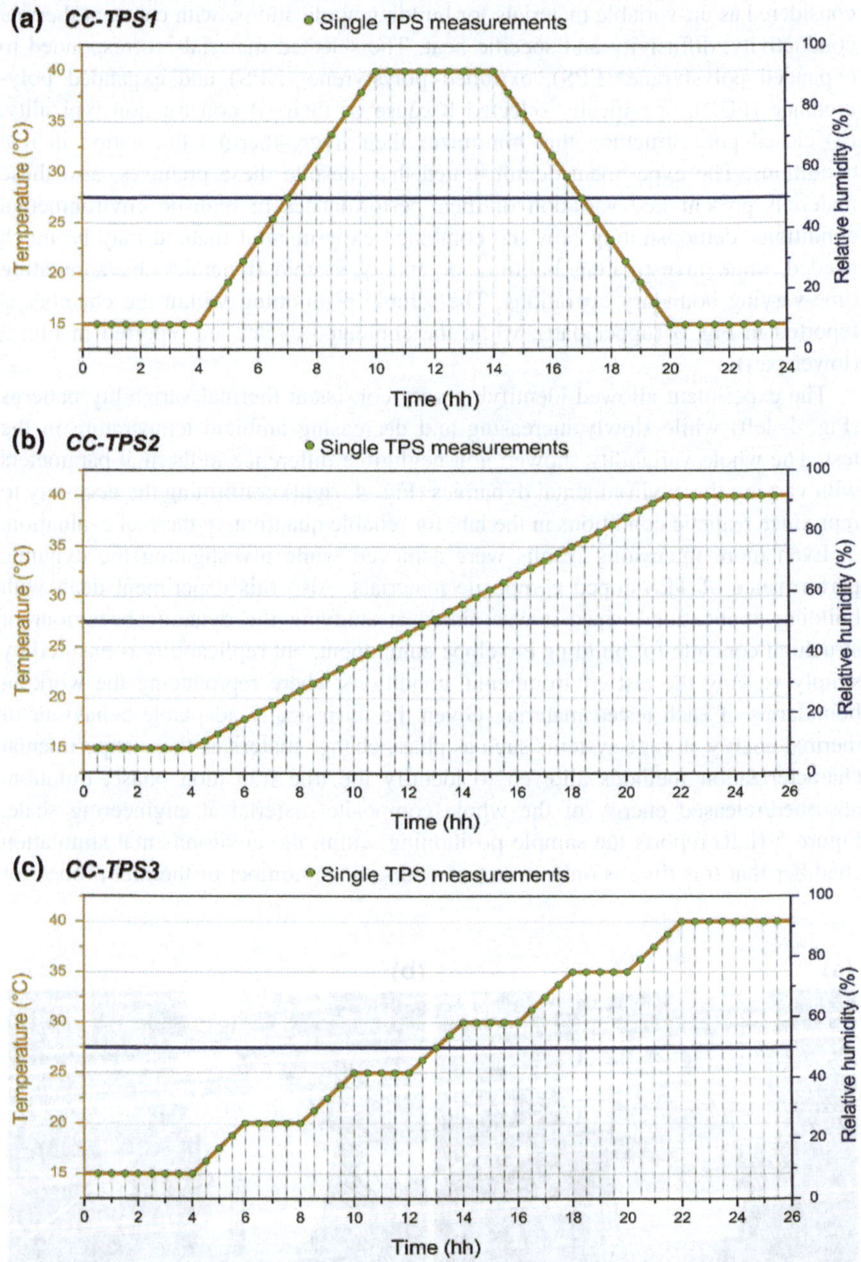

Fig. 3 (continued)

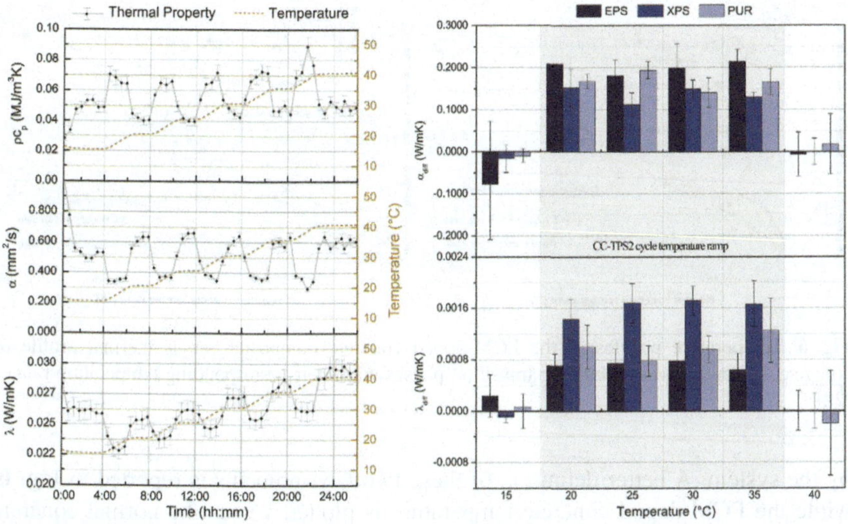

Fig. 4 Thermal conductivity, diffusivity and specific heat of insulation panel while increasing temperature step by step (left). Comparative variability of such parameters with varying environmental test dynamic conditions (right) [21]

detailed thermal characterization of concrete composite slabs in several positions, by assuming the heat transfer happening only through one free face of each slab, while all the others were thermally isolated. Figure 5 (right) shows the thermal fluctuation of such composite materials with increasing the PCM concentration within the concrete mix design, from 0 to 5%, during a 40-h long cycle. We can therefore identify the PCM activation during both melting and solidification, but also the duration of such TES effect and the time delay of the thermal wave detected

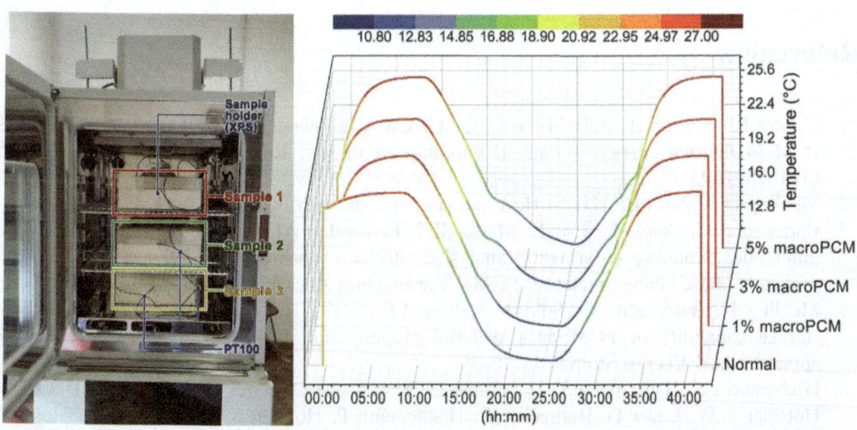

Fig. 5 Positioning of the composite samples in the test chamber (left), thermal profiles of the composites during the realistic thermal fluctuation (right) [22]

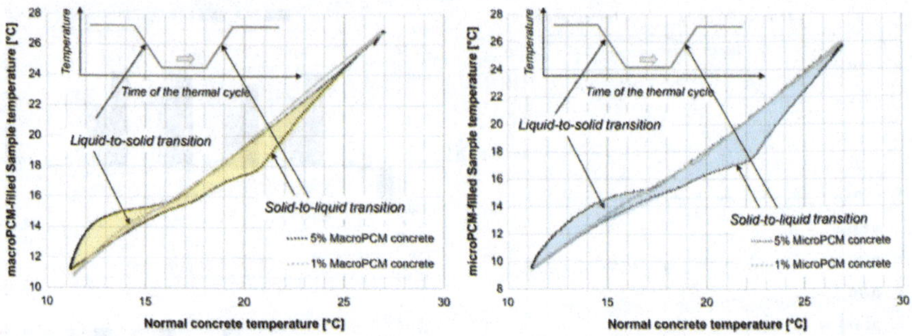

Fig. 6 Temperature profiles of the PCM-doped concrete composite versus thermal profile of concrete without PCM highlighting transition phases with their corresponding temperature ranges [22]

by the system. A better definition of these two key moments is reported in Fig. 6, while the PCM-doped concrete temperature is plotted versus the normal concrete without PCM. Here we may visualize the temperature range while the transition phase happens, and this information allows a designer to properly select the PCM inclusion technique, quantity, etc. by considering the real thermal dynamics of the composite, not only concerning the PCM itself, which is typically investigated by means of DSC and other techniques mentioned in this chapter.

4 PCM at Pilot Plant Scale

PCM can also be analysed at pilot plant scale. Information can be found in Chap. 11.

References

1. Cabeza LF, Castell A, Barreneche C, De Gracia A, Fernández AI (2011) Materials used as PCM in thermal energy storage in buildings: a review. Renew Sustain Energy Rev 15 (3):1675–1695
2. Mehling H, Cabeza LF (2008) Heat and cold storage with PCM. Springer, Berlin
3. Barreneche C, Solé A, Miró L, Martorell I, Fernández AI, Cabeza LF (2013) Study on differential scanning calorimetry analysis with two operation modes and organic and inorganic phase change material (PCM). Thermochim Acta 553:23–26
4. Mehling H, Barreneche C, Solé A, Cabeza LF (2017) The connection between the heat storage capability of PCM as a material property and their performance in real scale applications. J Energy Storage 13:35–39
5. Gschwander S, Haussmann T, Hagelstein G, Barreneche C, Ferrer G, Cabeza LF, Diarce G, Hohenauer W, Lager D, Rathgeber C, Hennemann P, Höhlein S, Mehling H, Peñalosa C, Lazaro A (2016) Standardization of pcm characterization via DSC. Refrig Sci Technol 70–75

6. Günther E, Hiebler S, Mehling H, Redlich R (2009) Enthalpy of phase change materials as a function of temperature: required accuracy and suitable measurement methods. Int J Thermophys 30(4):1257–1269
7. Ferrer G, Barreneche C, Solé A, Martorell I, Cabeza LF (2017) New proposed methodology for specific heat capacity determination of materials for thermal energy storage (TES) by DSC. J Energy Storage 11:1–6
8. Bauer T, Laing D, Tamme R (2012) Characterization of sodium nitrate as phase change material. Int J Thermophys 33(1):91–104
9. Sari A, Karaipekli A (2007) Thermal conductivity and latent heat thermal energy storage characteristics of paraffin/expanded graphite composite as phase change material. Appl Thermal Eng 27(8–9):1271–1277
10. Cheng W-L, Zhang R-M, Xie K, Liu N, Wang J (2010) Heat conduction enhanced shape-stabilized paraffin/HDPE composite PCMs by graphite addition: preparation and thermal properties. Sol Energy Mater Sol Cells 94(10):1636–1642
11. Ashby MF, Shercliff H, Cebon D (2007) Materials : engineering, science, processing and design. Elsevier, Butterworth-Heinemann. Callister WD, Rethwisch DG (2015) Fundamentals of materials science and engineering: an integrated approach, 7th edn. Wiley, New York
12. Miró L, Barreneche C, Ferrer G, Solé A, Martorell I, Cabeza LF (2016) Health hazard, cycling and thermal stability as key parameters when selecting a suitable phase change material (PCM). Thermochim Acta 627–629:39–47
13. Barreneche C, Gil A, Sheth F, Inés Fernández A, Cabeza LF (2013) Effect of d-mannitol polymorphism in its thermal energy storage capacity when it is used as PCM. Sol Energy 94:344–351
14. Fernández AI, Solé A, Giró-Paloma J, Martínez M, Hadjieva M, Boudenne A, Constantinescu M, Anghel EM, Malikova M, Krupa I, Peñalosa C, Lázaro A, Paksoy HO, Cellat K, Vecstaudža J, Bajare D. Sumiga B, Boh B, Haussmann T, Gschwander S, Weber R, Furmanski P, Jaworski M, Cabeza LF (2015) Unconventional experimental technologies used for phase change materials (PCM) characterization: Part 2—Morphological and structural characterization, physico-chemical stability and mechanical properties. Renew Sustain Energy Rev 43:1415–1426
15. Alkan C, Sari A, Karaipekli A, Uzun O (2009) Preparation, characterization, and thermal properties of microencapsulated phase change material for thermal energy storage. Sol Energy Mater Sol Cells 93(1):143–147
16. Gschwander S, Schossig P, Henning H-M (2005) Micro-encapsulated paraffin in phase-change slurries. Sol Energy Mater Sol Cells 89(2–3):307–315
17. Barreneche C, Solé A, Miró L, Martorell I, Fernández AI, Cabeza LF (2012) New methodology developed for the differential scanning calorimetry analysis of polymeric matrixes incorporating phase change material. Meas Sci Technol 23(8):085606
18. Gustavsson M, Gustafsson SE (2006) Thermal conductivity as an indicator of fat content in milk. Thermochim Acta 442(1–2):1–5
19. Gustavsson M, Hälldahl L (2006) Thermal conductivity measurements of thin insulating layers deposited on high-conducting sheets. Int J Thermophys 27(1):195–208
20. Sizov A, Cederkrantz D, Salmi L, Rosén A, Jacobson L, Gustafsson SE, Gustavsson M (2016) Thermal conductivity versus depth profiling of inhomogeneous materials using the hot disc technique. Rev Scientific Instrum 87(7):074901
21. Fabiani C, Pisello AL (2018) Coupling controlled environmental forcing and transient plane source method: an innovative thermal characterization procedure for building insulation materials. Appl Therm Eng 130:254–263
22. D'Alessandro A, Pisello AL, Fabiani C, Ubertini F, Cabeza LF, Cotana F (2018) Multifunctional smart concretes with novel phase change materials: mechanical and thermo-energy investigation. Appl Energy 212:1448–1461

Experimental Methods for the Characterization of Materials for Thermal Energy Storage with Chemical Reactions

Yukitaka Kato, Shigehiko Funayama, Elpida Piperopoulos and Candida Milone

Abstract The present chapter deals with the experimental characterisation methodologies for TES thermochemical materials with chemical reactions. In particular, thermogravimetric techniques, small-scale reactors configurations and methodologies for the evaluation of thermal energy and power density are discussed. Furthermore, morphological, structural and mechanical characterisations are introduced and applied to typical thermochemical TES materials.

1 Thermochemical Characterization of Thermochemical TES Materials by Thermogravimetric Analysis

A thermochemical energy storage material (TCM) is evaluated initially for thermochemical characterization by thermogravimetric analysis (TGA) with small sample (10–100 mg) in many cases. The gas–solid reaction with TCMs (e.g. metal hydroxides, metal carbonates, metal sulphates [1]) can be expressed by Eq. 8.1:

$$AB(s) \leftrightarrow A(s) + B(g) \tag{8.1}$$

Y. Kato (✉)
Laboratory for Advanced Nuclear Energy, Institute of Innovative Research, Tokyo Institute of Technology, 2-12-1-N1-22, O-Okayama, Meguro-Ku, Tokyo 152-8550, Japan
e-mail: yukitaka@lane.iir.titech.ac.jp

S. Funayama
Graduate Major in Nuclear Engineering, Tokyo Institute of Technology, 2-12-1-N1-22, Ookayama, Meguro-Ku, Tokyo 152-8550, Japan

E. Piperopoulos · C. Milone
Department of Engineering, University of Messina, Contrada di Dio, 98166 Messina, Italy

E. Piperopoulos · C. Milone
National Interuniversity Consortium of Materials Science and Technology (INSTM), 50121 Florence, Italy

A. Frazzica and L. F. Cabeza (eds.), *Recent Advancements in Materials and Systems for Thermal Energy Storage*, Green Energy and Technology,
https://doi.org/10.1007/978-3-319-96640-3_8

Applicability of a gas–solid reaction to thermochemical TES is studied in terms of reaction reversibility and cycle stability at first. Reversibility of TCM of candidate reactions can be examined from repetitive forward and backward reactions. The direction of a gas–solid reaction is determined by the sign of the change in the Gibbs free energy of the reaction expressed by Eq. 8.2:

$$\Delta_r G(T, P) = \Delta_r H^0 - T\Delta_r S^0 + RT\ln\left(\frac{P_{B(g)}}{P^0}\right) \tag{8.2}$$

where $\Delta_r G$ [J mol^{-1}], $\Delta_r H^0$ [J mol^{-1}], $\Delta_r S^0$ [J mol^{-1} K^{-1}], R [J mol^{-1} K^{-1}], T [K], $P_{B(g)}$ [kPa], P^0 [kPa] indicate the Gibbs free energy change, reaction enthalpy, reaction entropy, gas constant, temperature, the pressure of gas substance B in Eq. 8.1, standard pressure, respectively.

A forward reaction of Eq. 8.1 occurs on the condition that $\Delta_r G < 0$ while a backward reaction occurs on the condition that $\Delta_r G > 0$. The condition that $\Delta_r G = 0$ indicates chemical equilibrium. As can be seen from the Eq. 8.2, the sign of the Gibbs energy change, i.e. the direction of the reaction, depends on two variables namely, temperature and pressure. Therefore, we have to select proper conditions of temperature and pressure that can turn the sign of the Gibbs energy change to perform repetitive reaction experiment.

Cycling test of the reaction by TGA is often conducted by temperature swing operation at a constant pressure or by pressure swing operation at a constant temperature. Figure 8.1 shows a schematic diagram of TGA apparatus for reaction systems with water vapour as a gaseous reactant.

The temperature of the sample cell is measured by a thermocouple, which supports the sample cell. The temperature of the sample cell can be controlled at high heating or cooling rates by an infrared lamp heater inside the furnace and a chiller connected to the furnace.

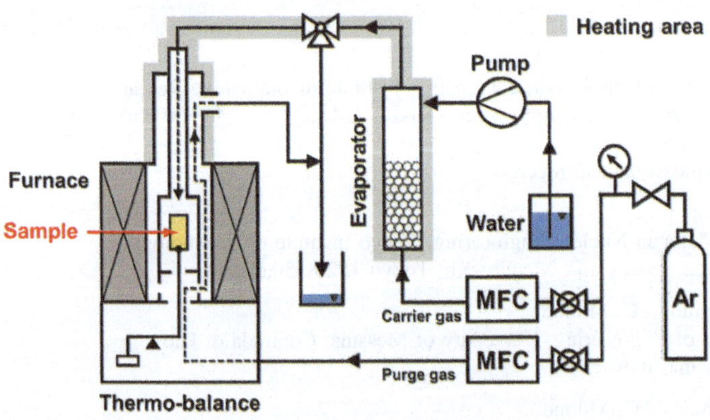

Fig. 8.1 Apparatus of thermogravimetric analysis (TGA) for water vapour reaction system

Partial pressure of water vapour in mixed gas of water vapour and carrier gas of argon is adjusted to a certain pressure (Fig. 8.1). However, closed type TGA apparatus controls the absolute pressure of pure atmosphere of gaseous reactant directly [2].

Reaction conversion, x[–], is calculated by Eq. 8.3 using a mass change of the sample detected in TGA:

$$x = 1 + \frac{\Delta m / M_B}{m_0 / M_{AB}} \tag{8.3}$$

where, Δm [g], m_0 [g], M_B [g mol^{-1}], M_{AB} [g mol^{-1}] denote mass change, initial mass of AB, molar mass of B, molar mass of AB, respectively.

Calcium oxide/calcium hydroxide/water ($CaO/Ca(OH)_2/H_2O$) reaction system is one of conventional reaction systems for thermochemical TES. The reaction formula is described in Eq. 8.4 [3].

$$Ca(OH)_2(s) \leftrightarrow CaO(s) + H_2O(g), \Delta_r H = 104 \text{ kJ mol}^{-1} \tag{8.4}$$

The reaction system is a promising candidate for thermochemical TES due to low material cost and high reactivity. The reaction is a reversible gas–solid reaction. A forward reaction is dehydration, which is an endothermic reaction, and a backward reaction is hydration, which is an exothermic reaction. The endothermic reaction and exothermic reactions are used for thermal storage process and thermal output process, respectively. Figure 8.2 shows an example of cycling test of $CaO/Ca(OH)_2/H_2O$ reaction system performed by TGA using temperature swing method. Conversion change from $x = 1$ to $x = 0$ shows $Ca(OH)_2$ dehydration (heat storage operation), and the change from $x = 0$ to $x = 1$ indicates CaO hydration (heat output operation). Enough durability to repetitive reactions in these 5 cycles can be confirmed from the figure.

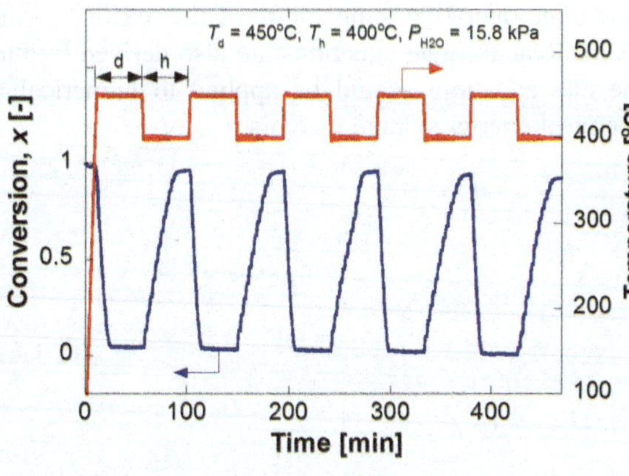

Fig. 8.2 Cycling test of $CaO/Ca(OH)_2$ by TGA (d: dehydration, h: hydration)

TCM analysis is conducted after the cycling test using TGA. Scanning Electron Microscopy (SEM) can be used to observe morphological difference between the material prior to cycling test and the material that experienced repetitive reaction. X-Ray diffraction is also used to examine products of side reactions and the change of crystallite size. This TCM analysis can support the material behaviour in terms of conversion rate or attained conversion observed in TGA.

Equilibrium measurements are also performed by TGA to obtain chemical equilibrium line. From the measured equilibrium line, we can identify the equilibrium temperature at a given pressure. The equilibrium temperature of a reaction at around atmospheric pressure (1–10^3 kPa) determines the range of operating temperature of the thermochemical TES using the reaction system. Reaction enthalpy is calculated from the slope of chemical equilibrium line and the van't Hoff equation [4]; reaction enthalpy is used to calculate energy density of the TCM.

Kinetic study of a gas–solid reaction for thermochemical storage is investigated by measurement of reaction conversion at isothermal and isobaric conditions using TGA. Reaction rate equations of the reaction can be derived by kinetic study. The rate equations are vital for numerical analysis of laboratory-scale reactors because numerical analysis of laboratory-scale reactors requires coupled equations, e.g. mass, energy balance equations and rate equations [5–7]. Therefore, TGA of a TCM contributes to the establishment of a numerical model of up-scaled reactors using the TCM and allow estimating thermochemical performances of the reactors.

TGA is also used to develop composite materials for thermochemical energy storage. The aim of the development of the composite materials is to enhance reactivity [8], increase conversion rate [9–11] or modify onset temperature of reaction [12]. It is important to investigate side reactions during repetitive reaction between storage material and combine materials for development of composite material with high durability. The durability of TCM can be examined by cycling test performed by TGA as well as in the development of the reaction systems.

In conclusion, TGA is a mandatory method to evaluate TCM performances and essential to establish a numerical model of up-scaled reactors. Applicability of a gas–solid reaction or a composite material for TCM is evaluated by TGA in terms of reversibility and cycle stability. TGA also identifies equilibrium properties which determine operating temperature of the reaction system and energy density of the TCM. Reaction rate equations are also derived by kinetic analysis using TGA and the rate equations would be applied to numerical analysis of practical thermochemical energy storage reactors.

2 Thermochemical Characterization of TCM Reactor at Laboratory Scale

TCM has to be evaluated by a packed bed reactor at laboratory scale in terms of energy density, thermal power and durability. Gas–solid reaction system is one of major choices for TCM. Powder state of TCM is not suitable for a practical reactor because powder state induces material scattering, bed material sintering and channelling and gas diffusion resistance increase, bed thermal conductivity decrease, and finally packed bed thermal performance degradation. Then, some shaped bulk materials of TCM in such as pellet, tablet and block figures are used in a bed reactor [13]. The bulk material reactor bed is expected to have high material density, high reaction gas diffusivity and high thermal conductivity for achieving high-performance TCM reactor with high energy density and high thermal power. Practical reactor requires high thermal power TCM to enhance the performance of the reactor and reduce its cost. Material durability of bulk TCM is a key performance for practical application where many reaction cycles are performed. Material analysis of the bulk TCM during cycle operation is required previously for TCM reactor evaluation. In general, TCM has intrinsic property of volume change during gas–solid reaction, which could affect contact thermal resistance between the material and heat exchanger. Thus, mechanical properties (e.g. stress distribution which affect sintering and breakage of pelletized bulk TCM) in bed reactor should be also investigated.

Experimental of TCM bed reactors in laboratory scale is important to have practical information of a design of a thermochemical TES system. Numerical analysis of the bed reactor by introducing a numerical model is useful for validation of thermal storage/output processes in the bed, and establishment of efficient thermochemical energy storage system with shorter period and lower cost development. To establish a numerical model, we have to consider coupled governing equations of heat transfer, mass transport and chemical reaction. Based on the numerical analysis we can calculate TES performances (e.g. reaction rate per unit volume or mass of the reactor) of an up-scaled TCM reactor. A proper numerical model also contributes to extract a limiting step during thermal storage/output process and the model could be used to estimate required transport properties of TCM as well as reactor design for an application.

An experimental study of thermochemical characterization of a TCM for $CaO/Ca(OH)_2/H_2O$ reaction system (Eq. 8.4) at laboratory-scale packed bed reactor is shown in Fig. 8.3 as an experimental example.

$Ca(OH)_2$ material is charged in the packed bed reactor installed inside the reaction chamber which is a vacuum vessel. The packed bed reactor is encapsulated in the reaction chamber, which is connected to the water reservoir. The water vapour generated in the reactor during dehydration condenses in the water reservoir. On the other hand, the water vapour for hydration is generated in the reservoir. The mass change of the chamber, induced by the transfer of the vapour during the reaction, is measured using an electronic balance installed under the chamber. This

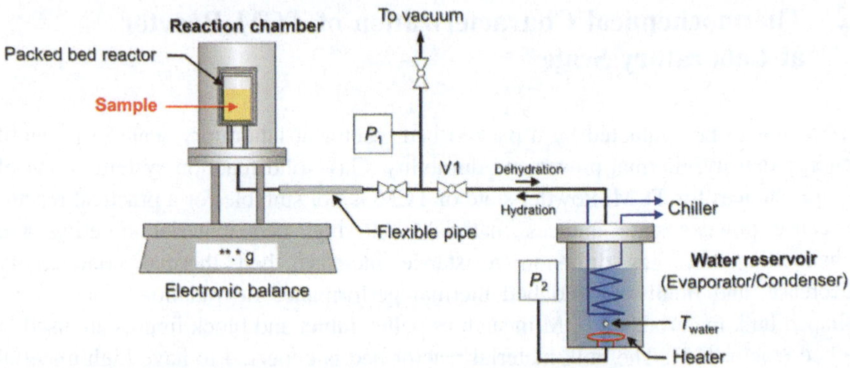

Fig. 8.3 Schematic diagram of a laboratory-scale reactor for thermochemical energy storage

Fig. 8.4 Sectional view of the TCM packed bed reactor

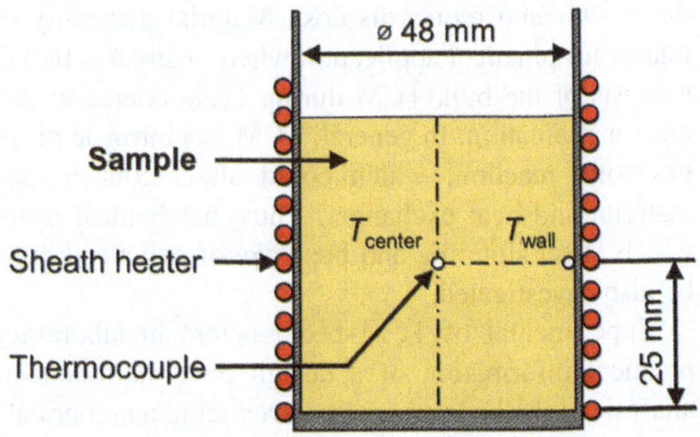

measurement system allows to auto-continuous water mass change during reaction with high-precision of 0.1 g order. The conversion, x [–], is calculated through Eq. 8.5 using the mass change measured at each cycle:

$$x = 1 + \frac{\Delta m / M_{H_2O}}{m_0 / M_{Ca(OH)_2}} \tag{8.5}$$

where Δm [g], m_0 [g], M_{H_2O}[g mol^{-1}], $M_{Ca(OH)_2}$ [g mol^{-1}] denote mass change of a sample, the initial mass of $Ca(OH)_2$, and molar mass of H_2O and $Ca(OH)_2$, respectively.

Figure 8.4 shows a sectional view of a packed bed reactor. Some thermocouples are installed in the reactor to measure the bed temperatures during a dehydration and hydration reaction and a sheath heater is placed around the reactor. The temperatures at the centre and on the inner wall of the reactor are measured during the reaction; the wall temperature is adjusted by a temperature controlling system. The temperature of water in the reservoir (T_{water}) is measured, and kept constant using a chiller during dehydration experiment and a heater during hydration experiment.

To fill the system by the pure atmosphere of water vapour, the air is evacuated from the system after charging a material in the reactor. The pressure inside the system corresponds to a saturated vapour pressure at T_{water} in the reservoir; thus, the pressure of the system is adjusted by T_{water} control.

Figure 8.5 shows measurement examples of temporal changes of temperature and conversion of a bed of calcium hydroxide pellets during dehydration and hydration. The dehydration experiment is initiated by increasing the wall temperature (T_{wall}) which is higher than the equilibrium temperature [14] corresponding to the pressure of water vapour (Fig. 8.5a). The temperature at the centre of the reactor (T_{center}) shows a typical endothermic curve, which has a period where the temperature remains constant.

Figure 8.5b shows the temperatures and reaction conversion of the bed during hydration. The experiment is initiated by opening the valve (V1, see Fig. 8.3) at $t = 0$ min, upon which water vapour is introduced into the reactor. The bed temperatures increase immediately after introducing the water vapour into the reactor. The maximum temperature of T_{center} agrees with the equilibrium temperature at the water vapour pressure

Hydration conversion rate is enhanced through a higher water vapour pressure due to the kinetics of the reaction. A higher vapour pressure leads to a higher attained temperature because of the higher equilibrium temperature; the attained temperature of the bed can be predictable by chemical equilibrium line of the reaction. The reactive stability of a TCM bed is demonstrated in terms of the rate of the reaction and total conversion.

In recent years, the $CaO/Ca(OH)_2/H_2O$ reaction system has been investigated experimentally at laboratory scale [15–21]. Agglomeration effect and channelling effect of powder bed has been reported. Therefore, there remains a need in terms of high-performance TCM development for practical application.

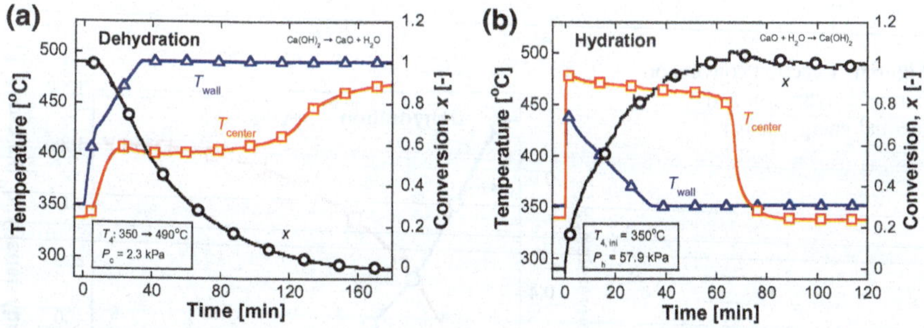

Fig. 8.5 Temperature and conversion of the TCM reactor bed: **a** dehydration and **b** hydration

3 Thermal Characterization of TCM

Thermal energy density and thermal storage/output power of TCMs are important properties to estimate the size and capability of a practical thermochemical TES reactor applied to conventional energy processes. This section shows evaluation methodology of thermal energy density and thermal storage/output powers of the TCM bed reactor for $CaO/Ca(OH)_2/H_2O$ system introduced in paragraph 2.

The apparatus for thermochemical storage discussed in paragraph 2 measures mass change of solid reactant/product of the gas–solid reaction and reaction conversion, x[–], is calculated by Eq. 8.5 using the mass change. Thermal energy density during dehydration, q_d [kJ m^{-3}] can be defined using conversion as follows:

$$q_d = \Delta_r H \frac{m_0}{M_{Ca(OH)_2}} \frac{1}{V_{bed}} (1 - x) \quad (8.6)$$

where V_{bed} [m^3] indicates the volume of an apparent TCM packed reactor bed.

Figure 8.6 shows reaction conversion and thermal energy density of the bed of calcium hydroxide pellets. As seen in Fig. 8.6, line of q has an inverted form of that of x, because q is in proportion to x with a constant value of $(\Delta_r H\, m_0) \cdot \left(M_{Ca(OH)_2}\, V_{bed}\right)^{-1}$ in Eq. 8.6. Regarding the TCM bed density of constant value, $m_0 \cdot V_{bed}^{-1}$, depends on the shape of TCM (such as powder, pellet, tablet, block, etc.) and kind of additive. High thermal energy storage density is favourable for TCM bed, so high $m_0 \cdot V_{bed}^{-1}$ is ideal. However, it affects x, and x directly affects q. Sometimes high density material caused low reaction gas diffusivity and small change of x. Reaction change of x is affected not only by $m_0 \cdot V_{bed}^{-1}$ but also conditions of reaction operation. Therefore, higher value of $m_0 \cdot V_{bed}^{-1}$ is required to be considered for development and operation of TCM. q is an important term for quantitative evaluation of TCM.

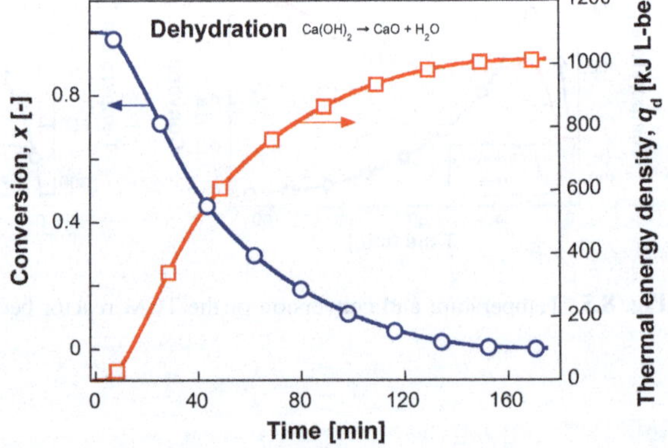

Fig. 8.6 Reaction conversion during dehydration and thermal energy density

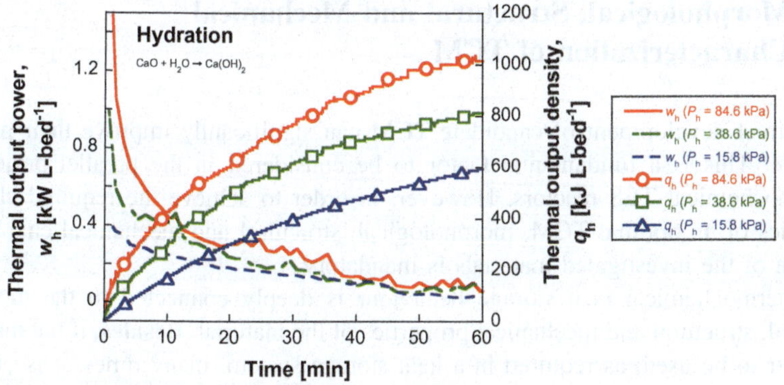

Fig. 8.7 Thermal output power and thermal output density at different pressures

Thermal output density during hydration, q_h, [kJ m^{-3}] can be defined as Eq. 8.7. Experimental results of q at different pressures are shown in Fig. 8.7,

$$q_h = \Delta_r H \frac{m_0}{M_{Ca(OH)_2}} \frac{1}{V_{bed}} x \tag{8.7}$$

Time derivatives of q_d and q_h are defined as thermal storage/output powers, w [kW/L-bed], as shown in Eq. 8.8. Thermal storage/output power are a key performance indicator for a thermochemical TES. Figure 8.7 shows also w_h.

$$w = \frac{dq}{dt} \tag{8.8}$$

As shown in Fig. 8.7, increasing water vapour pressure improves thermal output power and achieved density of thermal output for a hydration time. Thermal energy density and thermal output power of a bed of $Ca(OH)_2$ pellet TCM were evaluated around 1.0 MJ L-bed^{-1} and 1.0 kW L-bed^{-1}, respectively.

Evaluation of thermal energy density and thermal storage/output power of a TCM reactor at laboratory scale are important performance indicators for reduction of the volume of TCM reactor and enhancement of thermal performance of a thermochemical energy storage system. Thermal storage/output power depends on TCM and also depend on reactor design. Thus, development of TCM and reactor design should be studied simultaneously for development of practical storage system.

4 Morphological, Structural and Mechanical Characterization of TCM

A correct development of candidate TCM can significantly improve their performance. This is a fundamental factor to be considered in the parallel project of thermochemical TES reactors. However, in order to achieve the required characteristics of an optimal TCM, morphological, structural and mechanical characterization of the investigated materials is mandatory.

Thermochemical heat storage behaviour is deeply connected to the morphological, structural and mechanical properties of the material. Besides, if the material is ever to be used, as required in a heat storage system, many times, it is of vital importance that the reaction is reversible and can be performed a number of times as higher as possible, in other words the material should be cyclable. It has been observed that TCMs are not always stable (physically, mechanically and/or chemically) over several cycles of repeated reversible reactions. In some cases, the structural and morphological changes during the dehydration and hydration processes bring to improvements in the intrinsic kinetic properties of the interface grain reaction (nucleation/growth process) and the water vapour diffusion inside the grains [22, 23] and so in the thermochemical performances. These include volume changes due to the compaction and the expansion of the material and plastic behaviour (annealing, deformation), leading to the formation and/or removal of structural defects (grain boundaries, dislocations, porosity, lattice defects) [22]. As in the case of a pseudo-morphic structure that, appearing during the dehydration of salt hydrates in appropriate conditions [24] facilitates vapour transport through the crystal, thereby influencing the transport in time. In other cases these material changes can also lead to degradation (powdering, low intergrain heat transfer) and failure (powder bed packing, unreactive grain areas) of the storage material [22] negatively influencing the behaviour of the cycles following the first one [25, 26]. In addition, toxicity is related to the TCM cyclability. In fact, one of the required criteria for a possible implementation of an open sorption system in the residential sector is the non-toxic properties over cycles. Toxicity is for example, the most critical aspect for salt ammoniates employment in an individual house [22]. Besides, low stability under hydrothermal conditions can lead to decomposition of the salts (e.g. chlorides hydrolysis [27], carbonation of Na_2S [28]) reaching the formation of strong acid vapours (HCl for chloride salts and H_2S for Na_2S), which implicate corrosion problems and pollution. For the reasons described above the structural, morphological and mechanical characterization of cycled materials is mandatory for a complete investigation of the material and its possible application.

The most used characterization technique to investigate the structural properties of a TCM is X-Ray diffraction (XRD). This characterization procedure permits to generate by XRD scattering with the electron cloud of the investigated material, a unique diffractogram, characteristic of the material. The identification of the position (2ϑ angle) and the intensity (counts) of the XRD, characteristics of the lattice planes (hkl) of the atomic material structure, is the principle of the XRD data processing. During the analysis of a TCM, identifying a kinetic model, it is essential to include information about solid changes in textures, structures and bonding. For example, when MgO and CaO react with water vapour, their crystalline structure changes from cubic to hexagonal, the crystallite size is estimated by Scherrer equation analysing the peak width from the material's diffractogram [29]. The crystal structure changes and the coordination of water molecules with the studied material, during water vapour absorption, is related to the enthalpy value of the reaction which is related to the storage capacity of the technology implementing TCM [30]. Moreover, in order to confirm the thermal analysis study, XRD technique can be conducted in temperature, analysing in situ the dehydration of TC material, coupling each XRD pattern with a temperature of the dehydration profile [22, 31]. Finally yet importantly, during cycling of the thermochemical storage system, the TCM, changing its crystalline structure, can affect mechanical stability. In fact, this phenomenon may be accompanied by a change in the volume of the material. Due to the confinement of the material in the heat exchanger, a substantial volumetric increase or decrease may have a negative effect on contact between the material and the heat exchanger surfaces. Besides the resulting stress on the crystals may result in brittleness and powder forming, visible in measuring the height of the bed of the TCM during cycling [23].

In addition microstructural changes on the grain level are observed by Scanning Electron Microscopy (SEM). The surface of the TCM grains before and after TC reaction can be observed and this information is useful to see how kinetics is governed by, and if there is degradation. For example, microstructural investigations of metal oxide powders after three reduction/re-oxidation cycles show high sintering effects of the high temperature (900 °C), which leads to decreasing reaction rates with increasing number of cycles (Fig. 8.8) [25, 26].

Furthermore, the surface area and the pore distribution information are extremely useful to investigate heat transfer, mass transfer and diffusion. Surface area, in fact describes the exchange surface available during solid–gas reactions, influencing heat and mass transfer; while diffusion is affected by the particle size, the shape, the presence or the absence of impurities and porosity [30]. Usually the packing of the material during cycling brings to a decrease in surface area and in pore volume, what negatively affects the efficiency of the material. Lin et al. showed that CaO hydration rate decreases when increasing number of hydration cycles, since crystal grows and specific surface area reduces with the cycling progression [32]. The specific surface area is determined by nitrogen adsorption-desorption analysis at 77 K applying the Brunauer–Emmett–Teller (BET) method and pore distribution is obtained by a Barrett-Joyner-Halenda (BJH) analysis. Moreover, the use of the environmental SEM (ESEM) which allows the examination of specimens surrounded by a gaseous environment, also water vapour, during hydration rate may help the study of salt hydrates or solid–gas phase reactions of TCM [22], visually analysing the morphology changes of the investigated TC material during the reaction process. To corroborate SEM and XRD analysis, transmission microscopy (TEM) and Dynamic Light Scattering (DLS) are sometimes used and particle size distribution is obtained by statistical analysis [33, 34].

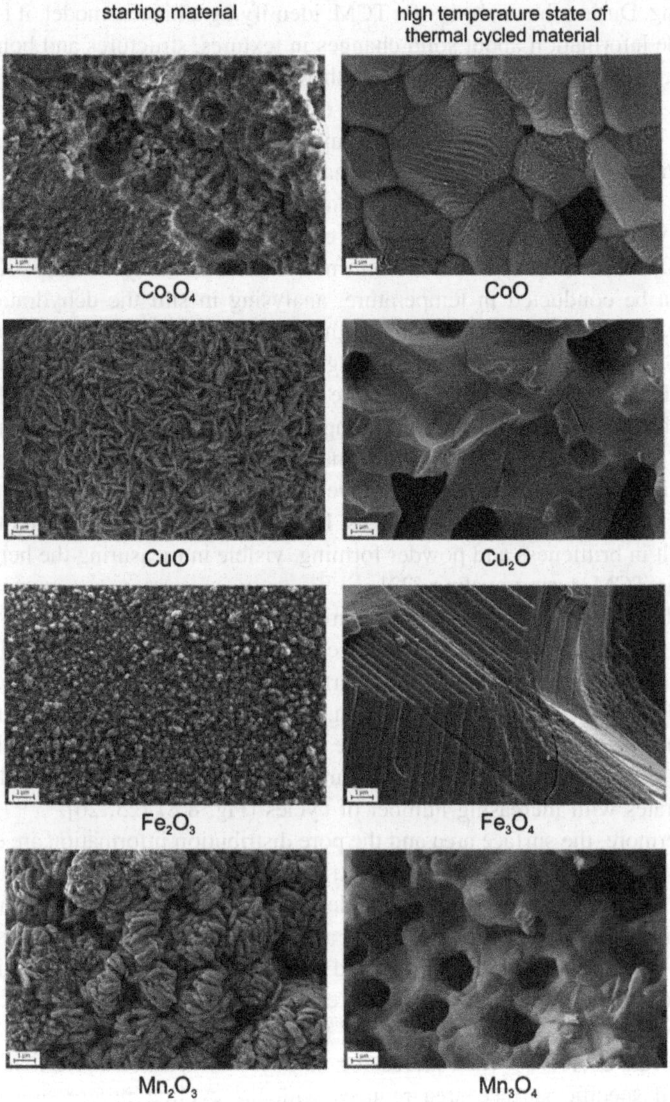

Fig. 8.8 SEM images of metal oxides, starting materials and high temperature state after 3 redox cycles [26]

Other important parameters to be studied are the physical properties and physical changes of TCM at the operation conditions. Bulk density is a useful information for calculating the energy density and consequently the encumbrance of the hypothetical accumulation system and it is generally larger in systems with a larger volume difference between pre-reaction and post-reaction material's state [24].

Solubility, deliquescence and melting point are undesired effects of TCM. It has be seen that $CaCl_2 \cdot 2\ H_2O$ is highly soluble, while $Ca(OH)_2$ and $CaSO_4 \cdot 2H_2O$ are not soluble in water. The solid amount that works as heat absorbent is useful to obtain the calculated yield of the reactor in which is located. If the solid presents less solubility, the probability of achieving the expected yield of the reactor is higher [30]. Melting is an undesired effect because it slows down the reaction, destroys the crystal structure and straightens the formation of the amorphous part, which is unable to bind water, decreasing the material's cyclability [34]. In this case, in situ Raman spectroscopy is used to characterize the molecular arrangement in material that cannot be identified with XRD measurements and so to analyse any chemical and physical modifications [35]. Furthermore, melting reduces the bed porosity of the material and thereby the vapour transport through the material bed, limiting the ability of the material to take up the water again [34].

Concluding, all the characterization techniques listed above are complementary. An incomplete study on solid reactants and products could lead to unexpected reactor yield. If a proper and complete characterization is achieved, the reliability of the suggested TC system with the proper TCM will increase.

References

1. André L, Abanades S, Flamant G (2016) Screening of thermochemical systems based on solid-gas reversible reactions for high temperature solar thermal energy storage. Renew Sustain Energ Rev 64:703–715. https://doi.org/10.1016/j.rser.2016.06.043
2. Takasu H, Ryu J, Kato Y (2017) Application of lithium orthosilicate for high-temperature thermochemical energy storage. Appl Energ 193:74–83. https://doi.org/10.1016/j.apenergy.2017.02.029
3. Halstead PE, Moore AE (1957) 769. The thermal dissociation of calcium hydroxide. J Chem Soc 3873–3875. https://doi.org/10.1039/jr9570003873
4. Denbigh KG (1981) The principles of chemical equilibrium, 4th edn. Cambridge University Press, Cambridge
5. Nagel T, Shao H, Roßkopf C et al (2014) The influence of gas-solid reaction kinetics in models of thermochemical heat storage under monotonic and cyclic loading. Appl Energ 136:289–302. https://doi.org/10.1016/j.apenergy.2014.08.104
6. Wang W, Kolditz O, Nagel T (2017) Parallel finite element modelling of multi-physical processes in thermochemical energy storage devices. Appl Energy 185:1954–1964. https://doi.org/10.1016/j.apenergy.2016.03.053
7. Ranjha Q, Oztekin A (2017) Numerical analyses of three-dimensional fixed reaction bed for thermochemical energy storage. Renew Energy 111:825–835. https://doi.org/10.1016/j.renene.2017.04.062
8. Kim ST, Ryu J, Kato Y (2011) Reactivity enhancement of chemical materials used in packed bed reactor of chemical heat pump. Prog Nucl Energy 53:1027–1033. https://doi.org/10.1016/j.pnucene.2011.05.013
9. Myagmarjav O, Zamengo M, Ryu J, Kato Y (2015) Energy density enhancement of chemical heat storage material for magnesium oxide/water chemical heat pump. Appl Therm Eng 91:377–386. https://doi.org/10.1016/j.applthermaleng.2015.08.008

10. Kariya J, Ryu J, Kato Y (2015) Reaction performance of calcium hydroxide and expanded graphite composites for chemical heat storage applications. ISIJ Int 55:457–463. https://doi.org/10.2355/isijinternational.55.457
11. Mastronardo E, Bonaccorsi L, Kato Y et al (2016) Thermochemical performance of carbon nanotubes based hybrid materials for $MgO/H_2O/Mg(OH)_2$ chemical heat pumps. Appl Energy 181:232–243. https://doi.org/10.1016/j.apenergy.2016.08.041
12. Shkatulov A, Aristov Y (2015) Modification of magnesium and calcium hydroxides with salts: an efficient way to advanced materials for storage of middle-temperature heat. Energy 85:667–676. https://doi.org/10.1016/J.ENERGY.2015.04.004
13. Zamengo M, Ryu J, Kato Y (2014) Thermochemical performance of magnesium hydroxide-expanded graphite pellets for chemical heat pump. Appl Therm Eng 64:339–347. https://doi.org/10.1016/j.applthermaleng.2013.12.036
14. Schaube F, Koch L, Wörner A, Müller-Steinhagen H (2012) A thermodynamic and kinetic study of the de-and rehydration of $Ca(OH)_2$ at high H_2O partial pressures for thermo-chemical heat storage. Thermochim Acta 538:9–20. https://doi.org/10.1016/j.tca.2012.03.003
15. Schaube F, Kohzer A, Schütz J et al (2013) De-and rehydration of $Ca(OH)_2$ in a reactor with direct heat transfer for thermo-chemical heat storage. Part A: experimental results. Chem Eng Res Des 91:856–864. https://doi.org/10.1016/j.cherd.2013.02.019
16. Pardo P, Anxionnaz-Minvielle Z, Rougé S et al (2014) $Ca(OH)_2/CaO$ reversible reaction in a fluidized bed reactor for thermochemical heat storage. Sol Energy 107:605–616. https://doi.org/10.1016/j.solener.2014.06.010
17. Roßkopf C, Haas M, Faik A et al (2014) Improving powder bed properties for thermochemical storage by adding nanoparticles. Energy Convers Manag 86:93–98. https://doi.org/10.1016/j.enconman.2014.05.017
18. Schmidt M, Szczukowski C, Roßkopf C et al (2014) Experimental results of a 10 kW high temperature thermochemical storage reactor based on calcium hydroxide. Appl Therm Eng 62:553–559. https://doi.org/10.1016/j.applthermaleng.2013.09.020
19. Yan J, Zhao CY (2016) Experimental study of $CaO/Ca(OH)_2$ in a fixed-bed reactor for thermochemical heat storage. Appl Energy 175:277–284. https://doi.org/10.1016/j.apenergy.2016.05.038
20. Kuwata K, Esaki T, Iwase D et al (2017) Long-term durability and reactivation of thermochemical heat storage driven by the $CaO/Ca(OH)_2$ reversible reaction. J Mater Sci Chem Eng 5:23–32. https://doi.org/10.4236/msce.2017.511003
21. Schmidt M, Linder M (2017) Power generation based on the $Ca(OH)_2$/ CaO thermochemical storage system—experimental investigation of discharge operation modes in lab scale and corresponding conceptual process design. Appl Energy 203:594–607. https://doi.org/10.1016/j.apenergy.2017.06.063
22. Ferchaud C (2016) Experimental study of salt hydrates for thermochemical seasonal heat storage. Technische Universiteit Eindhoven, Eindhoven
23. Roelands M, Cuypers R, Kruit KD et al (2015) Preparation & characterization of sodium sulfide hydrates for application in thermochemical storage systems. Energy Procedia 70:257–266. https://doi.org/10.1016/J.EGYPRO.2015.02.122
24. Donkers PAJ, Pel L, Adan OCG (2016) Experimental studies for the cyclability of salt hydrates for thermochemical heat storage. J Energy Storage 5:25–32. https://doi.org/10.1016/j.est.2015.11.005
25. Block T, Knoblauch N, Schmücker M (2014) The cobalt-oxide/iron-oxide binary system for use as high temperature thermochemical energy storage material. Thermochim Acta 577:25–32. https://doi.org/10.1016/J.TCA.2013.11.025
26. Block T, Schmücker M (2016) Metal oxides for thermochemical energy storage: a comparison of several metal oxide systems. Sol Energy 126:195–207. https://doi.org/10.1016/J.SOLENER.2015.12.032
27. Zondag H, Kikkert B, Smeding S et al (2013) Prototype thermochemical heat storage with open reactor system. Appl Energy 109:360–365. https://doi.org/10.1016/j.apenergy.2013.01.082

28. De Jong A-J, Trausel F, Finck C et al (2014) Thermochemical heat storage—system design issues. Energy Procedia 48:309–319. https://doi.org/10.1016/j.egypro.2014.02.036
29. Shkatulov A, Ryu J, Kato Y, Aristov Y (2012) Composite material "$Mg(OH)_2$/vermiculite": a promising new candidate for storage of middle temperature heat. Energy 44:1028–1034. https://doi.org/10.1016/J.ENERGY.2012.04.045
30. Solé A, Fontanet X, Barreneche C et al (2013) Requirements to consider when choosing a thermochemical material for solar energy storage. Sol Energy 97:398–404. https://doi.org/10.1016/j.solener.2013.08.038
31. Ferchaud C, Zondag H, Veldhuis J, de Boer R (2012) Study of the reversible water vapour sorption process of $MgSO_4.7H_2O$ and $MgCl_2.6H_2$ O under the conditions of seasonal solar heat storage. J Phys Conf Ser 395:12069. https://doi.org/10.1088/1742-6596/395/1/012069
32. Lin S, Harada M, Suzuki Y, Hatano H (2006) CaO hydration rate at high temperature ($\sim$1023 K). https://doi.org/10.1021/ef050257o
33. Carrillo AJ, Serrano DP, Pizarro P, Coronado JM (2014) Thermochemical heat storage based on the Mn_2O_3/Mn_3O_4 redox couple: influence of the initial particle size on the morphological evolution and cyclability. J Mater Chem A 2:19435–19443. https://doi.org/10.1039/C4TA03409K
34. van Essen VM, Zondag HA, Gores JC et al (2009) Characterization of MgSO[sub 4] hydrate for thermochemical seasonal heat storage. J Sol Energy Eng 131:41014. https://doi.org/10.1115/1.4000275
35. Brotton SJ, Kaiser RI (2013) In Situ Raman spectroscopic study of gypsum ($CaSO_4{\cdot}2H_2O$) and Epsomite ($MgSO_4{\cdot}7H_2O$) Dehydration Utilizing an Ultrasonic Levitator. J Phys Chem Lett 4:669–673. https://doi.org/10.1021/jz301861a

Experimental Methods for the Characterization of Materials for Sorption Storage

Vincenza Brancato and Andrea Frazzica

Abstract The present chapter deals with the main experimental characterizations available to measure the performance of sorbent working pairs used in sorption TES. The experimental methods for the evaluation of equilibrium sorption curves and the models for the sorption equilibrium description are introduced. Furthermore, theoretical and experimental approaches for the heat of sorption evaluation are defined and some examples, reported in the literature, about the thermal storage capacity evaluation of different working pairs are reported.

1 Introduction

As already discussed in previous chapters, the sorption process is a spontaneous phenomenon of accumulation of a gas or a vapour at the solid surface onto the bulk phase. The main constituents that take part during this process are:

(1) Sorbent: a solid on which sorption occurs [usually it is a porous material with a large specific surface (m^2/g)].
(2) Sorbate: a substance (gas or vapour) which is sorbed on the solid surface of the sorbent, which in the TES field is commonly water vapour.

Usually, the sorbent/sorbate pair is defined as 'working pair', since the TES capability is directly connected to the interaction between the two components.

The sorption process is generally classified as either physisorption (physical adsorption) or chemisorption (chemical sorption). As defined by IUPAC [1]: physisorption involves electrostatic and Van der Waals interactions (forces) of the same kind as those responsible for the imperfection of real gases and the condensation vapours, and which do not involve a significant change in the electronic orbital patterns of the species involved. Vice versa, chemisorption is a sorption

V. Brancato (✉) · A. Frazzica
CNR-ITAE, Istituto di Tecnologie Avanzate per l'Energia "Nicola Giordano", Via Salita S. Lucia sopra contesse 5, 98126 Santa Lucia, Messina, Italy
e-mail: vincenza.brancato@itae.cnr.it

A. Frazzica and L. F. Cabeza (eds.), *Recent Advancements in Materials and Systems for Thermal Energy Storage*, Green Energy and Technology,
https://doi.org/10.1007/978-3-319-96640-3_9

process in which the forces involved are valence forces of the same kind as those operating in the formation of chemical compounds.

In the following, the experimental approaches that can be adopted to evaluate the TES performance of different working pairs are introduced and discussed.

2 Experimental Parameters for Evaluation of Sorption TES Performance

The main features of working pairs for TES applications were reported in previous chapters. A schematic classification of the sorption TES, as reported by Yu [2], is represented in Fig. 1. In the present chapter, the introduced experimental methods are mainly referred to the characterization of solid adsorption and weak chemical reactions. Nevertheless, if properly adapted, these methods can be extended also to the testing of other sorption TES categories.

The experimental characterisation of working pairs is usually employed to evaluate the achievable TES capacity (TSC) at material level, whose definition is reported in Eq. 1:

$$\mathrm{TSC} = \int_{w_{des}}^{w_{sor}} \Delta H_{sor} \mathrm{d}w \tag{1}$$

where, ΔH_{sor} (J/g_{wat}) represents the differential heat of sorption, while w_{sor} and w_{des} (g_{wat}/g_{sor}) are the maximum and minimum sorption capacity, of the working pair at the given working boundary condition, respectively.

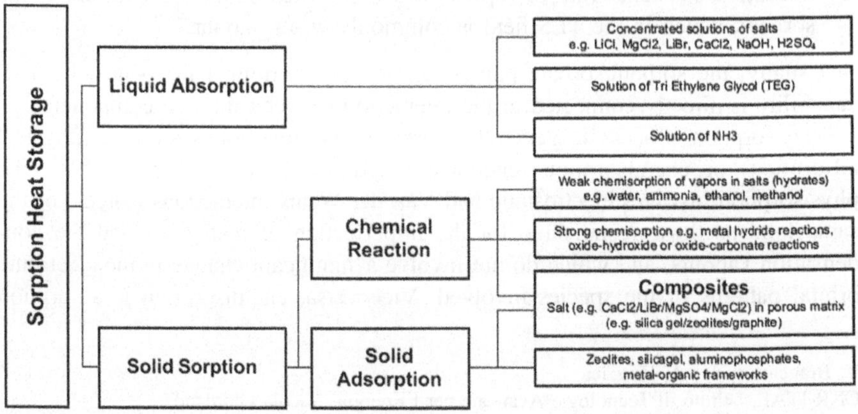

Fig. 1 Classification of sorption heat storage [3]

Accordingly, in order to obtain a complete material characterisation, the differential heat of sorption and the sorption equilibrium curves need to be evaluated.

Usually, the sorption equilibrium curves are experimentally evaluated by means of thermo-gravimetric or volumetric apparatuses, in which both temperature and water vapour pressure can be defined, to replicate the boundary conditions of a sorption TES. Accordingly, the maximum and minimum sorption capacity can be directly measured once fixed the working conditions. On the contrary, the differential heat of sorption can be either theoretically or experimentally derived. In the former case, it is obtained by applying equations of state able to describe the solid/vapour interaction and to derive the related heat of sorption, while in the latter it is evaluated by integrating the heat flow measured by differential scanning calorimetry (DSC) apparatuses.

The different experimental approaches are described in the following paragraphs.

3 Measurement of Equilibrium Sorption Curves

The measurement of equilibrium sorption curves represents one of the most important characterization methods, needed to evaluate the ability of the sorbent material to exchange sorbate under typical working conditions of a TES system.

Different parameters, such as equilibrium sorbate pressure, sorbent temperature, nature of the sorbent and the sorbate, highly influence the sorption capacity of the sorbent.

Accordingly, the sorption capacity, generally reported as sorption uptake, *w* (g/g), is a bi-variant function that, at the equilibrium, can be described as a function of two parameters, namely, temperature and pressure, as reported in Eq. 2:

$$w = f(T, p) \tag{2}$$

The uptake *w* is defined as the ratio between the sorbate loading over the dry sorbent mass, calculate at the equilibrium, see Eq. 3:

$$w(T, p) = \frac{m_{\text{sorbate}}}{m_{\text{dry sorbent}}} \tag{3}$$

The relationship between sorption uptake, pressure and temperature can be experimentally measured by means of three ways (Fig. 2) [4]:

- Sorption isotherm $w(P)_T$: it expresses the dependence between sorption uptake and pressure at constant temperature. Generally, it is observed a monotonous increasing of the sorbed amount with the pressure.
- Sorption isobar $w(T)_P$: it expresses the dependence between the sorbed amount and the temperature at constant pressure. It is characterized by the uptake decrease with temperature increase.

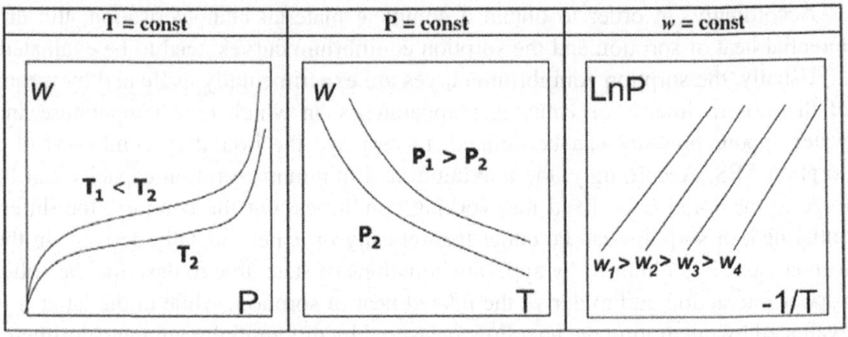

Fig. 2 Example of isotherms, isobars and isosteres [4]

- Sorption isostere $P(T)_w$: it expresses the dependence of the pressure on temperature at a constant uptake. Generally, the isosteres are represented in Clapeyron diagram, drawing ln P in dependence on $1/T$, and the curves becomes practically straight lines.

The sorption equilibrium curves are generally evaluated by two main procedures: the first one defined 'volumetric method' and the second one define 'gravimetric method'.

In the volumetric method, a known amount of gas/vapour is usually admitted into a confined and known volume containing the sorbent, maintained at constant temperature. When the sorption occurs, the pressure in the closed volume falls until the equilibrium is established. The amount of the sorbed gas/vapour at the equilibrium pressure is obtained as the difference between the amount of the sorbate admitted and the amount of gas necessary to completely fill the confined volume, usually considered as ideal gas. Therefore, the volumetric method is an indirect measurement method [5, 6].

Vice versa, the gravimetric method is a direct measure of the sorption equilibrium curves. In this case the change in weight of the sorbent during the sorption process, at a given temperature and pressure, is evaluated by means of a weight sensor (e.g. micro-balance). The sorbate pressure can be imposed either realizing a pure sorbate pressure inside the testing chamber, imposed by a thermo-regulated evaporator, or by fluxing a humidified stream over the tested sample, whose relative humidity is adjusted to reach the equivalent sorbate vapour pressure [7, 8].

Regardless the experimental apparatuses, so far a common procedure for measuring equilibrium curves was not agreed within the scientific community. Currently, the joint Annex 33/Task 58 promoted by the International Energy Agency (IEA) on the topic 'Material and Component Development for Thermal Energy Storage' is working on the definition of a common procedure, also by validating its application across different laboratories throughout Europe.

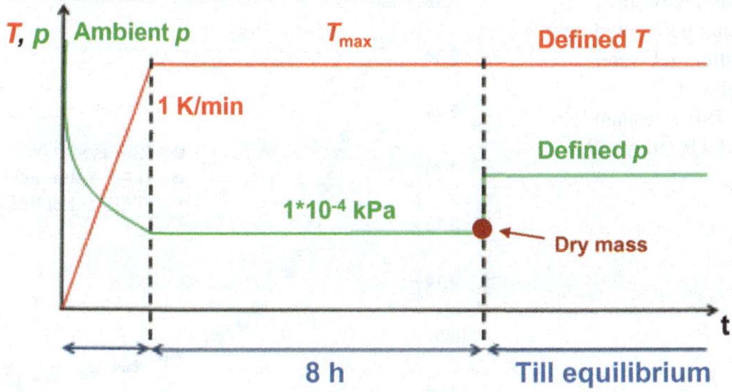

Fig. 3 Measurement procedure proposed in [9]

Recently, a first attempt in unifying the sorption characterization procedure was made by Henninger et al. [9], as outcome of the work performed during the Annex 34 of the IEA 'Thermally Driven Heat Pumps for Heating and Cooling'. The proposed procedure is depicted in Fig. 3.

The first step of the procedure is dedicated to the pre-treatment of the sorbent sample. During this phase, the material is slowly heated up until the maximum temperature and evacuated down to very low absolute pressure (e.g. 10^{-4} kPa as suggested). The identification of the maximum temperature depends on the nature of the sorbent. For instance, for very hydrophilic adsorbent materials, as zeolites 13X and 4A, it was suggested 300 °C. While for less hydrophilic materials, as silica gels, AlPOs and SAPOs, it was fixed at 150 °C. At least a continuous evacuation duration of 8 h was proposed. After this phase, the dry reference mass of the material is achieved. The next steps concern the measurement of the sorption equilibrium points of the working pair. The operating parameters of these steps depend on the experimental method employed, which means that they can be performed either in isothermal or isobaric mode. Depending on the level of detail that is requested, a certain number of equilibrium points can be taken during the measurement phase.

Figure 4 reports the comparison of measurements performed on a reference silica gel/water working pairs in two different laboratories, i.e. Fraunhofer ISE (Germany) and CNR-ITAE (Italy), applying the above-described methodology. In particular, isobaric measurements were performed by means of two different thermo-gravimetric apparatuses. The achieved results confirmed the reliability of the proposed methodology.

Based on this approach, once defined boundary operating conditions typical of sorption TES cycles, it is possible to evaluate the amount of sorbate exchanged per unit mass of sorbent material.

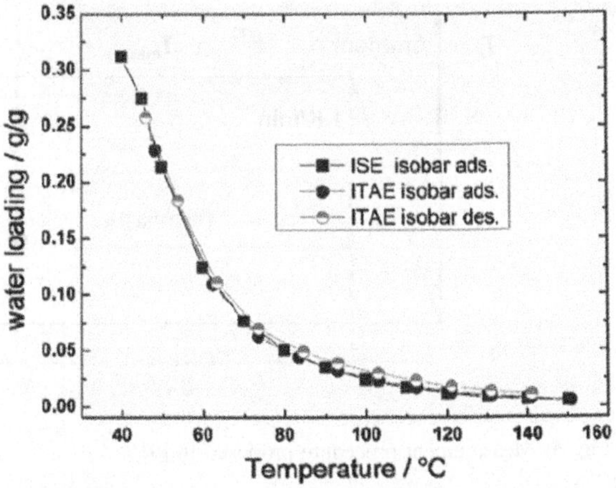

Fig. 4 Comparison of measurements performed on a reference silica gel/water working pairs at Fraunhofer ISE (Germany) and CNR-ITAE (Italy) [9]

4 Equilibrium Sorption Equations

As already introduced before, several theoretical and semi-empirical models were proposed in the literature to describe the sorption equilibrium occurring between the sorbate and the sorbent. These models allow to analytically evaluate the sorption capacity, once the working boundary conditions are defined. Furthermore, they can also allow the evaluation of heat of sorption, which represents the second parameter needed to evaluate the TSC. Several sorption equations take basis from the Langmuir theory (1918), the Gibbs approach (1980) and the Polanyi theory. In the present paragraph, the fundamental equations derived from these models are presented.

4.1 Langmuir Isotherm

One of the first theories about sorption equilibrium was proposed by Langmuir [10]. The Langmuir approach is based on a kinetic principle and it is founded on the assumption that, at the equilibrium, the rate of sorption is equal to the rate of desorption from the solid sorbent surface. The assumptions of Langmuir theory are three:

- Surface is homogeneous;
- Sorption on surface is localized at definite site;
- Each site can hold only one molecule.

The famous Langmuir equation, obtained starting from these assumptions and equating the rate of sorption and desorption is:

$$w(P,T) = w_0 \cdot \frac{b(T)P}{1+b(T)P} \quad (4)$$

where b is called affinity constant or Langmuir constant, and it is calculate by the following equation:

$$b = b_\infty \cdot \exp\left(\frac{Q}{RT}\right) \quad (5)$$

where, Q (kJ/mol) is the heat of sorption and is equal to the activation energy for desorption, which can be used for the evaluation of the TSC. While, b_∞ *is* a pre-exponential factor that is inversely proportional to the square root of the molecular weight of sorbate.

The affinity constant measures how strong is the attraction between the sorbate molecules with the sorbent surface. The trend of the Langmuir isotherm is highly influenced by the variation of Q, b and T, as evidenced in the following diagram (Fig. 5) [11].

Mainly:

- The sorbed amount increases when the Langmuir constant (b) or the heat of sorption (Q) increase. Indeed, there is a stronger affinity between sorbent and sorbate (high value of 'b') or there is a the higher energy barrier that the sorbed molecules have to overcome to evaporate from the sorbent surface (high value of 'Q');
- The sorbed amount decreases when the temperature (T) increases, due to the greater energy acquired by the sorbed molecule to evaporate.

Usually, the Langmuir equation well describes the experimental data, especially in case of monolayer physical adsorption.

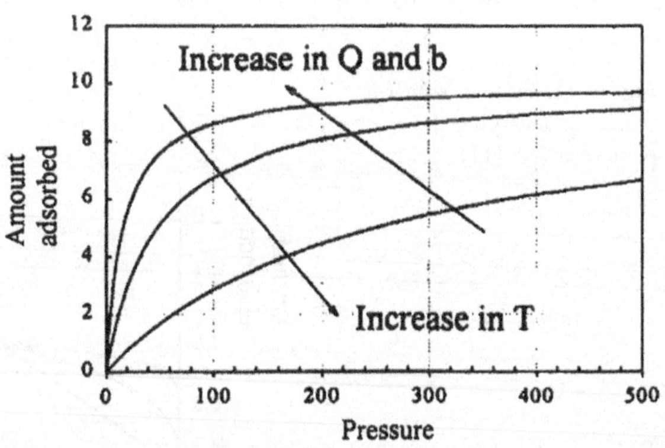

Fig. 5 Trend of Langmuir isotherm varying Q, b and T [11]

4.2 Semi-empirical Isotherm Equations

When the assumptions, on which the Langmuir is based, are not satisfied, other semi-empirical approaches can be used in order to describe the equilibrium curves. A common concept of these methods involves the idea of heterogeneity of the system, whether this heterogeneity derives from the solid sorbent, from the sorbate, or from a combination of both.

The earliest empirical equation was proposed by Freundlich [12]. The equation is written as:

$$w(P,T) = KP^{1/n} \tag{6}$$

where K and n are two parameters strongly dependent from the temperature. The parameter n is generally higher then unity. The higher is the value of n, the higher is the nonlinear behaviour of the isotherm, as evidenced from Fig. 6 [11].

This equation is widely used in the description of sorption of organics from aqueous streams onto activated carbons [11].

Since in the Freundlich equation the uptake increases continuously with the increase of the pressure, Sips [13] proposed a new corrected equation with a finite limit when the pressure is sufficiently high that is:

$$w(P,T) = w_0 \cdot \frac{(bP)^{1/n}}{1+(bP)^{1/n}} \tag{7}$$

This equation is similar to the Langmuir equation; the only difference is the additional parameter n, in the numerator of the fraction. When this parameter is equal to 1.0, the Langmuir equation is again obtained. Therefore, in this case, the parameter n is strongly linked to the system heterogeneity, indeed the higher is the parameter the higher is the heterogeneity of the system. Figure 7 shows the trend of the Sips equation by varying the parameter n. The behaviour of the isotherm

Fig. 6 Freundlich isotherms obtained at varying the parameter "n" [11]

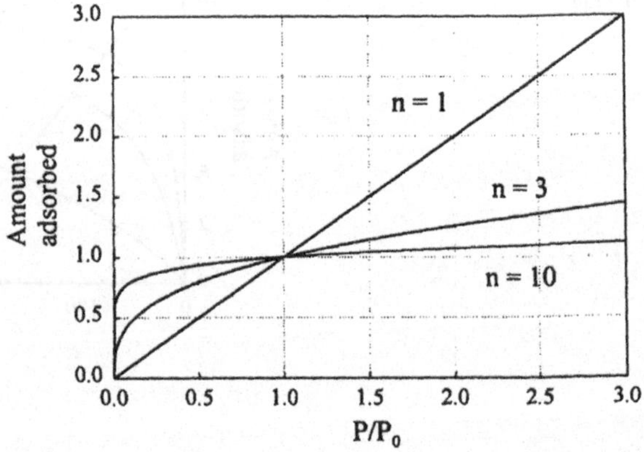

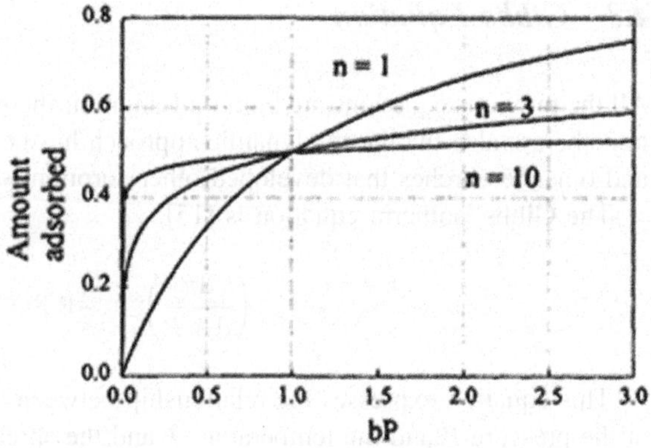

Fig. 7 Sips isotherms obtained at varying the parameter "*n*" [11]

follows the same trend of Langmuir curves except for the finite saturation at high pressure.

The Sips equation provides excellent result when it is implemented in the study of sorption of hydrocarbons on activated carbon.

Both Freundlich and Sisp equations do not overcome the disadvantage at low pressure where they do not go over to the linear dependence of the Henry law. For this reason, Toth [14] proposed the following equation:

$$w(P,T) = w_0 \cdot \frac{bP}{\left(1 + (bP)^t\right)^{1/t}} \tag{8}$$

where t is a parameter generally lower than the unity. Both b and t depend on the specific pair sorbent-sorbate. If t = 1.0, Toth isotherm is reduced to the basic Langmuir equation, therefore, as before mentioned for Sips equation, this parameter depends on the system heterogeneity. The higher is the deviation from the unity, the more heterogeneous is the system. The effect of the t parameter is shown in Fig. 8.

The Toth equation has correct limits when the pressure approaches either zero or infinity.

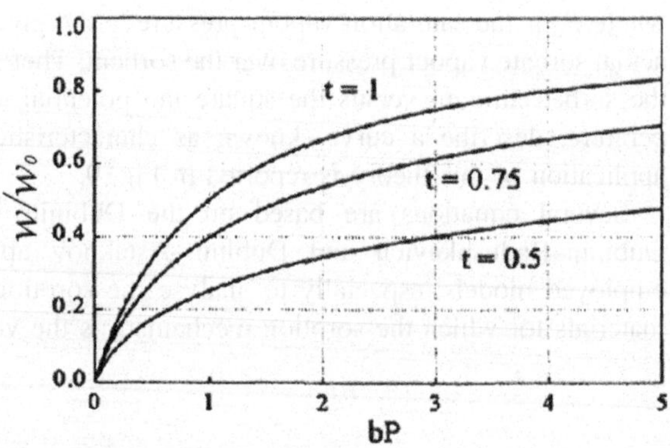

Fig. 8 Toth isotherms obtained at varying the parameter "*t*" [11]

4.3 Gibbs Equation

All the previous equations are built on Langmuir theory that has a kinetic approach; nonetheless also the thermodynamic approach has been taken in account by Gibbs and other researches that developed other sorption isotherm equations.

The Gibbs isotherm equation is [15]:

$$\left(\frac{\mathrm{d}\pi}{\mathrm{d}\ln P}\right)_T = \frac{n}{S}\mathrm{RT} \tag{9}$$

This equation expresses the relationship between the number n of sorbed moles at the pressure P and the temperature T and the spreading pressure (π), which, for the sorbed phase, is equivalent, from the thermodynamic point of view, to the pressure for a real fluid [16].

This equation can be transformed into various forms, since the correlation that it expresses is not very useful for presentation of sorption equilibrium data (i.e. linear isotherm describes by Henry equation and Volmer isotherm) [11].

4.4 Polanyi Empirical Theory of Sorption Equilibrium

All the above-presented methods are found on theoretical approach; nonetheless, also many empirical equations were proposed to depict the experimental data.

Polanyi potential theory is the approach on which is focused the popular Dubinin equation [17], highly used for describing experimental data of water vapour sorption onto microporous adsorbent materials. The key factor of the Dubinin equation is that the temperature dependence is entirely expressed in the sorption potential A (J/mol) defined as:

$$A = \mathrm{RT}\ln\left(\frac{P_0}{P}\right) \tag{10}$$

where P_0 is the saturation vapour pressure at the given temperature, while P is the actual sorbate vapour pressure over the sorbent. Therefore, plotting the logarithm of the sorbed amount versus the square the potential, all the data, at different temperature, describe a curve, known as characteristic curve. An example of the application of this theory is reported in Fig. 9.

Several equations are based on the Dubinin formalism, among them the Dubinin–Radushkevich and Dubinin–Astakhov approaches are still the most employed models, especially to analyse the sorption equilibrium of microporous materials for which the sorption mechanism is the volume filling of micropores.

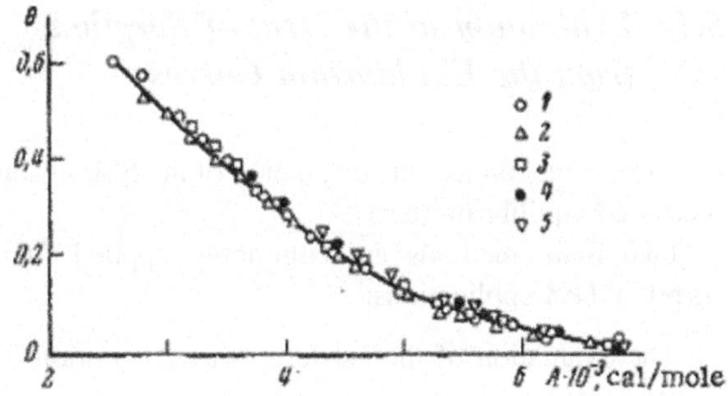

Fig. 9 Characteristic curve of the adsorption of carbon dioxide on LiX zeolite at: (1) 273 K; (2) 298 K; (3) 318 K; (4) 343 K; (5) 363 K published by Dubinin in 1971 [18]

The Dubinin–Radushkevich (D–R) equation is:

$$w = w_0 \exp\left[-\left(\frac{A}{E}\right)^2\right] = w_0 \exp\left[-\left(\frac{A}{\beta E_0}\right)^2\right] \tag{11}$$

its generalization by Astakhov (D–A) [18]:

$$w = w_0 \exp\left[-\left(\frac{A}{E}\right)^n\right] = w_0 \exp\left[-\left(\frac{A}{\beta E_0}\right)^n\right] \tag{12}$$

where w_0 (g/g) is the maximum sorbed volume, E_0 (J/mol) is the so-called characteristic energy, β is a constant that depends only on the sorbate nature, and n describes the surface heterogeneity.

The Dubinin–Astakhov equation still represents the most widely employed for describing sorption equilibrium of water vapour onto solid sorbents for TES applications. Indeed, it is able to describe the equilibrium by means of only one parameter, the sorption potential A, itself function of pressure and temperature. Furthermore, as will be described in the following chapter, it allows to directly determining the differential heat of sorption.

5 Heat of Sorption Calculation

Two different methods for the definition of the heat of sorption can be pursued. One is based on the exploitation of the equilibrium curves description introduced in the former chapter, which allows deriving the sorption heat of the given working pair. The other one is a fully experimental methods, which consists in the direct measurement of the heat flow during the sorption/desorption phase by means of a DSC apparatus, working under typical conditions of a TES.

In the following the two proposed approaches are detailed.

5.1 Evaluation of the Heat of Sorption from the Equilibrium Curves

As already mentioned, the evaluation of the heat of sorption can be derived from the measured equilibrium curves.

Two main methods are commonly applied for working pairs employed in sorption TES applications:

- The evaluation of the isosteric heat of sorption from the slope of the sorption isosteres;
- The derivation of the heat of sorption form the sorption potential defined in the Polanyi theory.

The isosteric heat of sorption derives from the van't Hoff equation and is defined as follows in Eq. 13 [7]:

$$\Delta H_{\mathrm{sor}} = R\frac{\mathrm{d}(\ln P)}{\mathrm{d}(1/T)}\bigg|_{w} \tag{13}$$

According to Eq. 13, the isosteric heat of sorption can be calculated by evaluating the slope of the each isoster when reported in the ln *P* versus 1/*T* plot. Once defined the boundary conditions and the relative operating pressures and temperatures, it is possible to integrate the Eq. 13, to obtain the integral isosteric heat of sorption, necessary to calculate the TSC, as defined in Eq. 1. The drawing of the isosteric chart can be performed starting from the equilibrium points measured under either isotherm or isobaric conditions. A direct method is represented by the experimental measurement of the isosters, as depicted in Fig. 2. Despite this approach can be considered more reliable, the complexity of experimental apparatuses for the accurate measurement of isosteres has hindered the diffusion of this methodology. Accordingly, often this calculation is graphically derived.

Starting from the description of the equilibrium points suggested by Polanyi and applied with the D–R and D–A approaches, it is possible to derive the differential heat of sorption according to Eq. 14:

$$\Delta H_{\mathrm{sor}} = L + A - T\alpha W\left(\frac{\partial \Delta A}{\partial w}\right)_{T} \tag{14}$$

where L (J/g) is the heat of evaporation of the sorbate (i.e. water), α (1/K) is the coefficient of thermal expansion of the and w (cm^3/g) is the sorbed volume. A (J/g) is the sorption potential, defined in Eq. 10. Also in this case, once the working conditions are defined, it can be integrated to calculate the TSC.

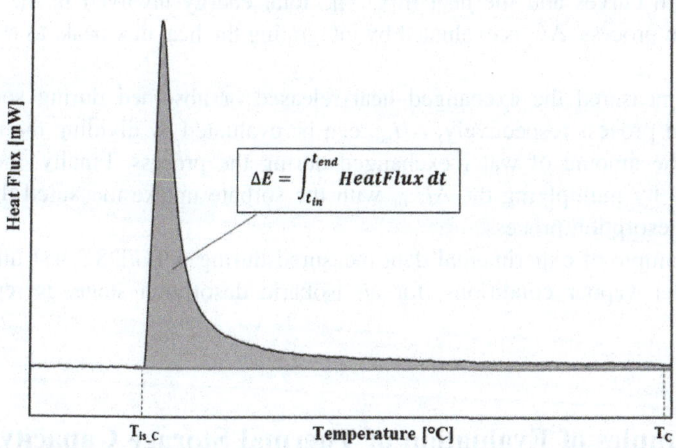

Fig. 10 Calculation of the heat of sortpion by integrating the heat flux measured by the DSC [19]

5.2 *Experimental Evaluation of the Heat of Sorption from DSC Experiments*

The heat of sorption can be measured experimentally by means of DSC apparatuses working under conditions typical of sorption TES. Recently, Frazzica et al. [19] proposed an approach based on a modified DSC/TG instrument, able to work under

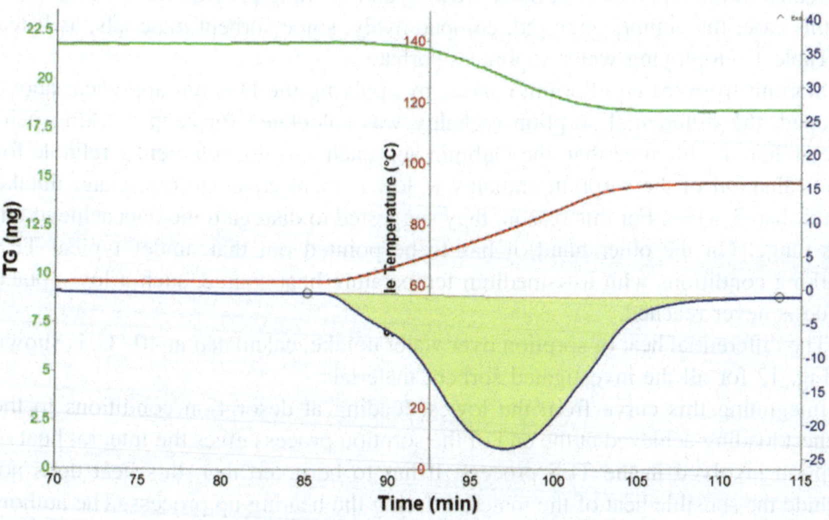

Fig. 11 Example of dynamic evolution of weight (green line), temperature (red line) and heat flux (blue line), under a typical isobaric desorption test [19]

saturated vapour conditions, thus contemporary measuring the sorption/desorption equilibrium curves and the heat flux. The total energy involved in the sorption/desorption process, ΔE, is evaluated by integrating the heat flux peak, as reported in Fig. 10.

Once measured the exchanged heat released or absorbed during sorption or desorption process respectively, ΔH_{sor} can be evaluated by dividing the measured heat by the amount of water exchanged during the process. Finally, the TSC is calculated by multiplying the ΔH_{sor} with the sorbate uptake measured during the sorption/desorption process.

An example of experimental data measured during a TG/DSC test under saturated water vapour conditions, for an isobaric desorption stage, is reported in Fig. 11.

6 Examples of Evaluation of Thermal Storage Capacity of Sorption Working Pairs

In the following paragraph, two examples of calculation of the TSC, performed by means of the different approaches above detailed, are reported.

6.1 Theoretical Approach

An example of the calculation performed according to the theoretical approach was presented in the report of IEA Solar Heating and Cooling programme Task 32 [20]. In this case, the authors analysed, comparatively, some sorbent materials, as listed in Table 1, employing water vapour as sorbate.

Starting from the equilibrium curves, by applying the Dubinin approach, above detailed, the differential sorption enthalpy was calculated for each working pair. The authors highlighted that, the Dubinin approach was not sufficiently reliable for the evaluation of the sorption enthalpy at low sorbent coverage (i.e. water uptake lower than 3 wt%). For this reason, they suggested to disregard the data achieved in this range. On the other hand, it has to be pointed out that, under typical TES working conditions with low-medium temperature heat source, such a low uptake value is never reached.

The differential heat of sorption over water uptake, calculated at 40 °C, is shown in Fig. 12 for all the investigated sorbent materials.

Integrating this curve from the lowest loading at desorption conditions to the highest loading achieved at the end of the sorption process gives the integral heat of sorption involved in the TES process. It has to be noted that, this heat does not include the sensible heat of the material during the heating up process. The authors have selected two operating conditions, which are:

Table 1 Selected materials investigated in IEA Task 32 [20]

Label	Company	Type
13X	Bayer	Zeolite 13X
NaCa 5A	Bayer	Zeolite 5A
SC Y 1/16	UOP	Zeolite Y
127B	Grace Davison	Microporous silica gel
LE-32	Engelhardt	Macroporous silica gel
SWS-1L	University of Novosibirsk	Mesoporous silica gel impregnated with $CaCl_2$
FAM-Z02	Mitsubishi Chemicals Corporation	Zeolite based molecular sieve

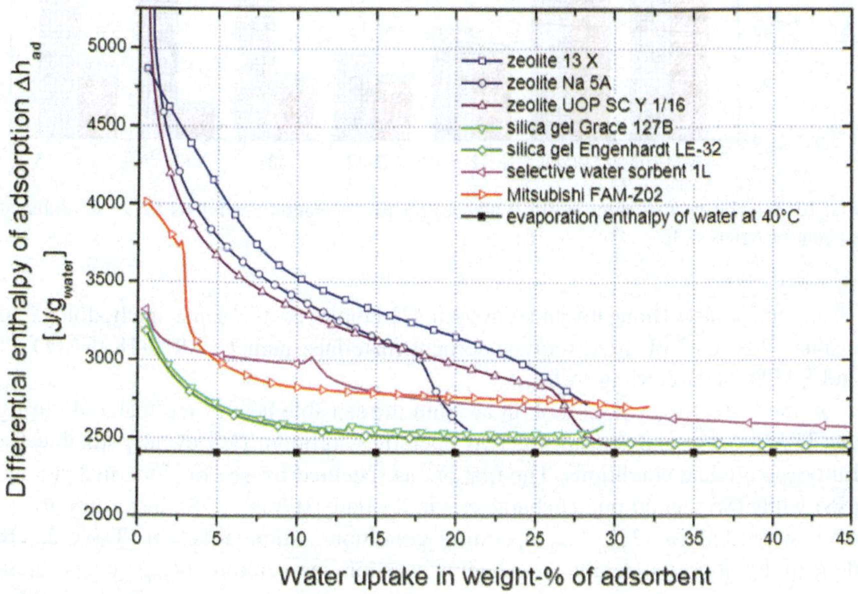

Fig. 12 Differential enthalpy of sorption as a function of water uptake [20]

1. 90 °C desorption temperature
   40 °C minimum sorption temperature and condensation temperature
   15 °C evaporation temperature
   label: 90/40/15 °C
2. 160 °C desorption temperature
   40 °C minimum sorption temperature and condensation temperature
   5 °C evaporation temperature
   label: 160/40/5 °C.

The obtained results are summarized in Fig. 13.

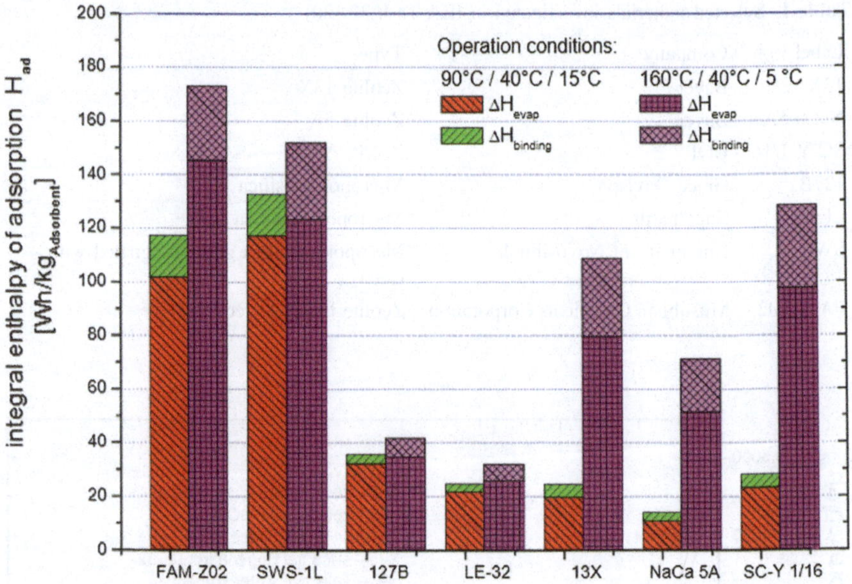

Fig. 13 Integral heat of sorption (in Wh/kg_{sor}) for two operation conditions for seven different sorbent materials [20]

Recently, also Brancato and Frazzica [7] employed the same methodology to evaluate the TSC of three zeotype sorbent materials, namely AlPO-18, FAPO-34 and SAPO-34, according to Eq. 1.

In this case, the authors took in account the sensible heat of the material during the TES process and considered two operating sorption TES cycles, suitable for European climate conditions. The first one is specified for seasonal thermal storage (SS) while the second one is characteristic for daily storage (DS), according to the ones suggested in [21]. The operating conditions, summarized in Table 2, are defined by four temperatures, namely, sorption temperature (T_{sor}), evaporation temperature (T_{ev}), desorption temperature (T_{des}) and condensation temperature (T_c).

Figure 14 shows the obtained comparison histogram of TSC for the zeotype materials for seasonal and daily storage varying the desorption temperature. The authors concluded that AlPO-18, FAPO-34, and SAPO-34 showed a good TSC and therefore these materials are suitable for application in sorption TES.

6.2 *Experimental Approach*

By means of the experimental device proposed by Frazzica et al. [19] and above described, Palomba et al. [22] investigated the sorption ability as well as the thermal TSC of a composite sorbent of water, based on LiCl/vermiculite, under two relevant

Table 2 Operating conditions for SS and DS

Cycle	T_{sor} (°C)	T_{ev} (°C)	T_{des} (°C)	T_c (°C)
SS	35	10	70, 75, 80, 85, 90	30
DS	35	5	70, 75, 80, 85, 90	30

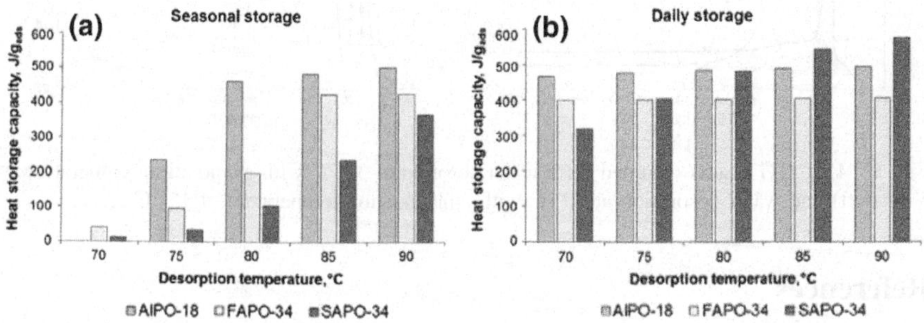

Fig. 14 TSC under typical **a** seasonal storage and **b** daily storage [7]

Table 3 Operating conditions of the DS and SS cycles

Cycle	T_{dis} (°C)	$T_{in.dis}$ (°C)	T_{ev} (°C)	T_{ch} (°C)	$T_{in.ch}$ (°C)	T_c (°C)
SS	35	62	10	80	54	30
DS	35	68	5	80	46	15

boundary conditions, namely, seasonal (SS) and daily (DS) storage applications. The cycle defined as Seasonal heat Storage (SS) can be considered applicable for regions where the heating demand can be, at least partially, compensated by the excess of solar heat stored in summer. While, the cycle defined as Daily heat Storage (DS) in winter, can be considered suitable for regions where the solar thermal energy during day-time is intense enough to be stored, and then used for heating at night, when the ambient temperature falls below the comfort level. The operating conditions of a TES cycle, namely, the temperatures of released heat (T_{dis}), evaporation (T_{ev}), regeneration (T_{ch}), and condensation (T_c) were selected in accordance with the climatic data in the regions where these cycles can be realized (Table 3) [21].

Figure 15 represents the DSC/TG curves obtained for the DS cycle. Similar curve were obtained for SS cycle. As evident, the diagrams show a typical evolution of heat flux, temperature and weight signals recorded during charging/discharging phases of water vapour on the LiCl/vermiculite composite.

The authors found that achievable TSC of this material are of 2.15 and 1.89 kJ/g for SS and DS cycle respectively.

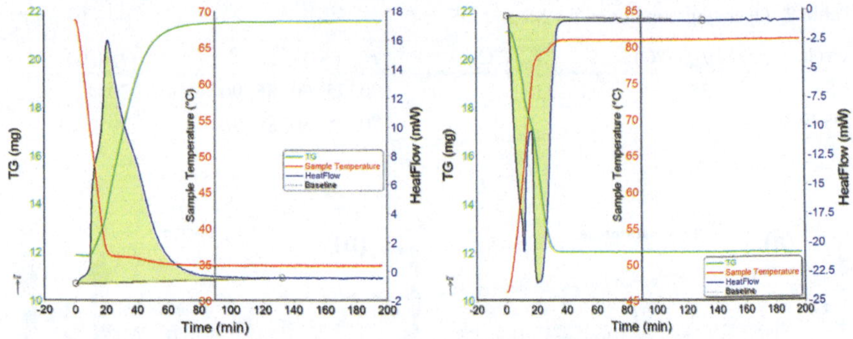

Fig. 15 DSC/TG results obtained for a water sorption at 35 °C with evaporation temperature of 5 °C (left) and water desorption at 80 °C with condensation temperature of 15 °C

References

1. Nič M (1997) IUPAC compendium of chemical terminology, 2nd edn. IUPAC Compend Chem Terminol Gold B 2. https://doi.org/10.1351/goldbook.f02510
2. Yu N, Wang RZ, Wang LW (2013) Sorption thermal storage for solar energy. Prog Energy Combust Sci 39:489–514. https://doi.org/10.1016/j.pecs.2013.05.004
3. Scapino L, Zondag HA, Van Bael J et al (2017) Sorption heat storage for long-term low-temperature applications: a review on the advancements at material and prototype scale. Appl Energy 190:920–948. https://doi.org/10.1016/j.apenergy.2016.12.148
4. Aristov YI (2012) Adsorptive transformation of heat: principles of construction of adsorbents database. Appl Therm Eng 42:18–24. https://doi.org/10.1016/j.applthermaleng.2011.02.024
5. Dawoud B, Aristov Y (2003) Experimental study on the kinetics of water vapor sorption on selective water sorbents, silica gel and alumina under typical operating conditions of sorption heat pumps. Int J Heat Mass Transf 46:273–281. https://doi.org/10.1016/S0017-9310(02)00288-0
6. Schnabel L, Tatlier M, Schmidt F, Erdem-Şenatalar A (2010) Adsorption kinetics of zeolite coatings directly crystallized on metal supports for heat pump applications (adsorption kinetics of zeolite coatings). Appl Therm Eng 30:1409–1416. https://doi.org/10.1016/J.APPLTHERMALENG.2010.02.030
7. Brancato V, Frazzica A (2018) Characterisation and comparative analysis of zeotype water adsorbents for heat transformation applications. Sol Energy Mater Sol Cells 180:91–102. https://doi.org/10.1016/J.SOLMAT.2018.02.035
8. Sapienza A, Velte A, Girnik I et al (2017) "Water - Silica Siogel" working pair for adsorption chillers: adsorption equilibrium and dynamics. Renew Energy 110. https://doi.org/10.1016/j.renene.2016.09.065
9. S.k. Henninger, A. Freni, L. Schnabel, P. Schossig GR (2011) Unified water adsoprtion measurement procedure for sorption materials. In: International sorption heat pump conference (ISHPC11), Padova, pp 513–522
10. Langmuir I (1916) The constitution and fundamental properties of solids and liquids. Part I Solids J Am Chem Soc 38:2221–2295. https://doi.org/10.1021/ja02268a002
11. Moradi O (2011) Thermodynamics of interfaces, thermodynamics - interaction studies - solids, liquids and gases. In: Dr. Juan Carlos Moreno PirajÃ¡n (ed) ISBN: 978-953-307-563-1, InTech, Available from: http://www.intechopen.com/books/thermodynamics-interaction-studies-solids-liquids-andgases/thermodynamics-of-interfaces

12. Freundlich H (1932) Of the adsorption of gases. Section II. Kinetics and energetics of gas adsorption. Introductory paper to section II. Trans Faraday Soc 28:195. https://doi.org/10.1039/tf9322800195
13. Sips R (1948) On the structure of a catalyst surface. J Chem Phys 16:490–495. https://doi.org/10.1063/1.1746922
14. J. T (1971) State equation of the solid-gas interface layers. Chim Acad Sci Hungaricae 63:311–328
15. Gibbs JW (1928) The collected works of J. W. Gibbs: Thermodynamics. Longmans, Green, co., London
16. Myers AL, Prausnitz JM (1965) Thermodynamics of mixed-gas adsorption. AIChE J 11:121–127. https://doi.org/10.1002/aic.690110125
17. Dubinin MM (1980) Water vapor adsorption and the microporous structures of carbonaceous adsorbents. Carbon N Y 18:355–364. https://doi.org/10.1016/0008-6223(80)90007-X
18. Dubinin MM, Astakhov VA (1971) Development of the concepts of volume filling of micropores in the adsorption of gases and vapors by microporous adsorbents—communication 2. General bases of the theory of adsorption of gases and vapors on zeolites. Bull Acad Sci USSR Div Chem Sci 20:8–12. https://doi.org/10.1007/BF00849308
19. Frazzica A, Sapienza A, Freni A (2014) Novel experimental methodology for the characterization of thermodynamic performance of advanced working pairs for adsorptive heat transformers. Appl Therm Eng 72:229–236. https://doi.org/10.1016/J.APPLTHERMALENG.2014.07.005
20. Bales C, Gantenbein P, Jaehnig D et al (2008) 031—chemical and sorption storage—results from IEA-SHC task 32. Eurosun 2008:1–8
21. Grekova A, Gordeeva L, Aristov Y (2016) Composite sorbents "Li/Ca halogenides inside multi-wall carbon nano-tubes" for thermal energy storage. Sol Energy Mater Sol Cells 155:176–183. https://doi.org/10.1016/j.solmat.2016.06.006
22. Palomba V, Gordeeva L, Brancato V et al (2017) Experimental characterization of a lab-scale adsorption thermal storage based on the LiCl/Vermiculite composite sorbent. In: International sorption heat pump conference, Tokyo

Corrosion Characterization in Components for Thermal Energy Storage Applications

Ángel G. Fernández, Margalida Fullana, Luigi Calabrese, Edoardo Proverbio and Luisa F. Cabeza

Abstract This chapter presents the corrosion characterisation methods used for thermal energy storage, in molten salts used in CSP plants and phase change materials (PCM) used for latent energy storage. The chapter also summarises results found in the literature and give hints on corrosion mitigation techniques.

1 Introduction

Concentrating solar power (CSP) technology captures and stores the sun energy in the form of heat, using materials that are low cost and materially stable for decades. Thus, CSP with thermal energy storage (TES) is an effective solution to the integration challenge, delivering renewable energy while providing important capacity, reliability, and stability to the grid and consequently enabling increased penetration of variable-generation renewable electricity technologies. Today most advanced CSP systems are towers integrated with two-tank molten salt TES, delivering thermal energy at 565 °C for integration with conventional steam-Rankine power cycles. This design has lowered the cost of CSP electricity by about 50% over the prior generation of parabolic trough systems; however, the decrease in cost of CSP

Á.G. Fernández
Energy Development Center, Universidad de Antofagasta,
Av. Univ Antofagasta, Antofagasta, Chile

M. Fullana · L. F. Cabeza (✉)
GREiA Research Group, INSPIRES Research Centre,
Universitat de Lleida, Pere de Cabrera s/n, 25001 Lleida, Spain
e-mail: lcabeza@diei.udl.cat

M. Fullana
CIRIAF—Interuniversity Research Centre on Pollution and Environment
"Mauro Felli", Via Duranti 63, 06125 Perugia, Italy

L. Calabrese · E. Proverbio
Department of Engineering, Universita degli Studi di Messina,
Contrada di Dio Sant'Agata, Messina, Italy

A. Frazzica and L. F. Cabeza (eds.), *Recent Advancements in Materials and Systems for Thermal Energy Storage*, Green Energy and Technology,
https://doi.org/10.1007/978-3-319-96640-3_10

technologies has not kept pace with the falling cost of solar photovoltaic (PV) systems and further investigation is needed [1].

The current state-of-the-art power tower designs use a nitrate molten salt mixture that together with high operation temperatures and the impurities present in the salts tend to drive corrosion. The accepted corrosion limits must be established to determine the corrosion mitigation approaches needed and the maintenance protocols required. Determining the cost of materials, cost of corrosion control, and lifetime of components can allow determining the cost of subsystems such as the receiver, TES, piping, pumps, valves, sensors, and heat tracing. High-temperature corrosion is one of the most important issues for materials selection, structural design, and service life prediction of engineering parts exposed to harsh environments. Corrosion prevention plays a critical role on reliability, quality control, safety, and profitability of any industrial sector associated with high-temperature processes.

The costs attributed to corrosion damage of all kinds have been estimated to be on the order of 3–5% of industrialized countries gross national product (GNP). But it is not clear exactly how much cost to attribute to specific problems (as molten salt pumping costs) [2].

2 Gravimetric and Electrochemical Concepts for Corrosion. Experimental Methods

2.1 Introduction

There are different evaluation techniques for a proper corrosion study that are summarized in Table 1.

Commonly used corrosion evaluation techniques involve mass gain or lost in the steel immersed in the corrosive environment and in the recent years more efforts have been made on the development of electrochemical techniques.

Table 1 Corrosion evaluation techniques

Inspection technique	Advantage	Disadvantage
Weight loss (WL)	Standard methodology	Requires long time of exposure
Weight gain (WG)	Simple, inexpensive	Requires long time of exposure
Weight loss (DGA)	Minimized human error	Low implementation
Linear sweep voltammetry (LSV)	Fast, sensitive, accurate	Destructive technique
Electrochemical impedance spectroscopy (EIS)	Fast, sensitive, accurate	Pseudo reference electrodes at high temperatures
Electrochemical noise (ECN)	Non-destructive	Complex analysis of data
Cyclic voltammetry	Quick identification of redox potentials	Dependent on time along

Electrochemical impedance spectroscopy (EIS) has gained popularity since it can be performed in situ, and also because it usually does not require any artificial acceleration of the corrosion process.

Linear Sweep Voltammetry (LSV) and Electrochemical Noise (ECN) techniques are focused on corrosion rate evaluation. LSV is a simple but destructive technique and on the other hand, using ECN technique, an in situ corrosion rate could be obtained. However, the ECN data analysis is complex since it involves the fluctuations of current and potential.

Finally, another electrochemical technique that could be applied in order to obtain a rapid identification of redox potentials is cyclic voltammetry. It can provide information about the thermodynamics of a redox process, kinetics of heterogeneous electron-transfer reactions and analysis of adsorption processes.

2.2 *Corrosion: Experimental Methods*

2.2.1 Gravimetric Method

The common method to evaluate the corrosion caused by immersion in a molten salt environment is the gravimetric technique. The specimens used in the corrosion test need to be dimensioned in order to know the area exposed to the molten salt attack. Once the samples are removed and the corrosion process has finished, coupons are cooled slowly and then dried and weighed. An average taken from five values of their weights is usually carried out [3, 4]. The formula used to calculate the mass gain over time is:

$$\frac{\Delta m}{S_0} = \frac{m_\mathrm{f} - m_\mathrm{i}}{S_0}, \tag{1}$$

where m_i is the initial mass of the specimen, m_f is the mass of the same specimen at time t, and S_0 is the initial area of the specimen. The gravimetric curve is completed using this value (Y axis) and time of exposure (X axis).

The main problem this technique presents is the spallation and loss of the corrosion layer, due to handling and to thermal expansion coefficient variation among the different oxides formed in the steel surface, producing errors in the gravimetric values. In order to avoid this uncertainty, the other common way to evaluate the gravimetric corrosion by means of the ASTM standard (G1-03) [5] involves the opposite evaluation.

This method analyses the steel weight loss removing the corrosion layer produced after the corrosion process. The methodology to evaluate the corrosion proposed by the ASTM standard (G1-03) has been widely used in many fields. The procedure involves corroded metal immersed in a cleaning solution that reacts with the oxide layer. Depending on the nature of the base material the ASTM standard proposes different solutions and thermal treatments in order to remove only corrosion products. To determine the mass loss of the base metal when removing corrosion products, replicated non-corroded control specimens should be cleaned by the same procedure being used as reference samples. During the process, manual brushing could be necessary as the results are dependent on the ability to remove this corrosion layers.

The metal weight loss technique can be complemented with an evaluation of the variation in the metal thickness (metal recession) after the corrosion test as well as the cleaning procedure following the ASTM standard. Dignam and Huggins [6] defined the metal as a thin film through which light is transmitted. The light used may be visible, ultraviolet or infrared; the wavelength used is one absorbed or reflected by the metal film. The attenuation of the light beam caused by reflection and absorption of the metal film is calibrated in terms of the metal film thickness. By this method, metal recession can be continuously measured to an accuracy of about ±10 A.

Gravimetric techniques (weight gain and loss) should be performed jointly for a proper development of the corrosion test. Initially, the study of mass gain is focused on the corrosion layers obtained (including XRD and SEM analysis) and after that, coupons must be cleaned with the corresponding cleaning solution in order to study the weight loss during the corrosion process.

A new methodology called Dynamic Gravimetric Analysis (DGA) [7] was proposed lately in order to minimize the handling of the sample. The samples are hung on a platinum wire (connected to a precision balance) and fully submerged into the solution recommended in the ASTM G1-03. The cleaning solution starts to attack the metal after removing all the metal oxide. This point is detected by a slope change in the mass loss versus time registered. Therefore, the corroded layer is eroded by the cleaning solution avoiding the handling.

2.2.2 The Electrochemical Cell

Most experimental methods aimed to investigate corrosion sensitivity of metals in specific environment are based on an electrochemical cell setup, equipped with a measurement instrument called potentiostat. A potentiostat is an electronic device able to controls voltage and current flow between two electrodes. Typically, a potentiostat works with an electrochemical cell containing three electrodes, according to the scheme reported in Fig. 1:

- Working Electrode: The electrochemical reactions under evaluation occur at the working electrode. In corrosion testing, the working electrode is a sample of the corroding metal. This is the analogous of coupons in weight loss testing. The working electrode can be bare or coated metal sample.
- Reference Electrode: A reference electrode is used for measuring the working electrode potential. A reference electrode should have a constant electrochemical potential and no current flows through it. Reference electrodes are standardized elements commercially available.

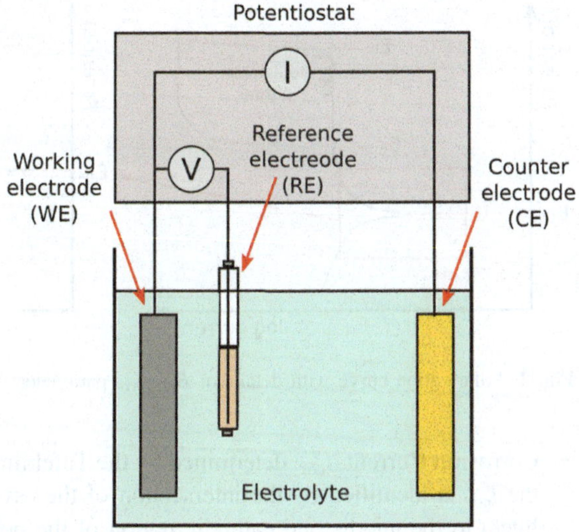

Fig. 1 Scheme of a three electrode electrochemical cell

- Counter Electrode: The counter electrode is a metallic or conductive element that completes the cell circuit. The counter electrode must be inert in the electrolyte, common materials used are platinum, activated titanium or graphite.

2.2.3 Potentiodynamic Polarisation

This technique can provide useful information on the working electrode corrosion rate, corrosion behaviour and it can be also used to assess the susceptibility of a material to corrosion in specific environmental conditions. Polarization methods are based on variation of the electrochemical potential of the working electrode and subsequent monitoring of the related current flow through the counter electrode. Next, some common potentiodynamic polarisation methods [8] are briefly introduced.

Anodic/Cathodic polarization: the potential is changed in the anodic (more positive direction) and/or cathodic direction (more negative direction) causing the working electrode to act as an anode or a cathode, respectively. In the former case, the polarization causes the metal dissolution and a negative charge flow (positive current) from it. In the latter case (negative current), oxygen or hydrogen are the main reduction reactions occurring on the metal surface. In Fig. 2 a scheme of anodic and cathodic polarization curve for an active material is reported.

In the figure criteria used to determine corrosion potential and current is shown:

- Corrosion Potential, E_{cor}: potential at which there is a negative cusp (lowest current value) in the polarization curve or potential at open circuit.

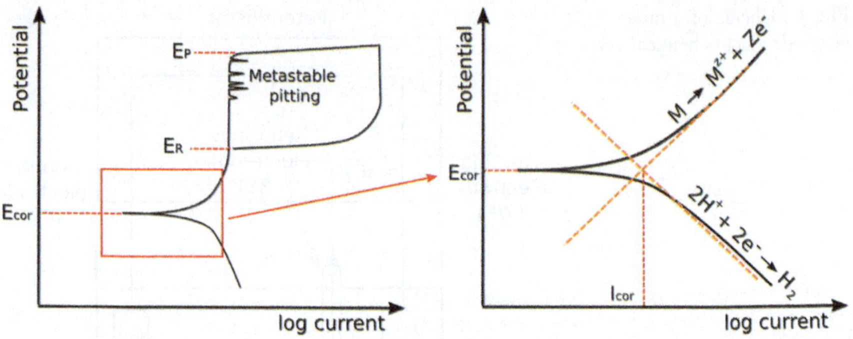

Fig. 2 Polarization curve with details of E_{cor}, i_{cor} parameters

- Corrosion Current, i_{cor}: determined by the Tafel line method, according to which the i_{cor} is identified by the intersection of the asymptotic lines (dashed orange lines) to the anodic and cathodic region of the polarization curve.

Linear polarization resistance (LPR): By using this technique, the polarization resistance of a material (working electrode) is defined as the slope of the potential-current density curve (both axis on linear scale) at the free corrosion potential. This measurement method, where not significantly perturbation of the WE takes place, is a not-destructive test.

Cyclic polarization test: This test is usually applied to assess pitting susceptibility of the working electrode. The potential is swept in a single cycle in upward and backward direction. The shape and magnitude of the hysteresis area gives information about localized corrosion sensitivity, based on the differences between the open circuit and the return passivation potential during the downward polarization. The existence of the hysteresis is usually indicative of pitting, while the size of the loop is often related to the intensity of pitting phenomenon.

2.2.4 Electrochemical Impedance Spectroscopy

Electrochemical impedance spectroscopy (EIS) is a non-stationary technique based on the differentiation of the reactive phenomena by their relaxation time [9].

The electrochemical impedance analysis is usually performed by applying an AC potential with a specific frequency (ω) to an electrochemical cell and then acquiring the corresponding current flowing through the cell. In particular applying a sinusoidal potential excitation ($V = V_o \sin(\omega t)$, the response will be a sinusoidal current signal, shifted in phase, φ ($I = I_o \sin(\omega t + \varphi)$).

If the sinusoidal voltage excitation is characterized by a low amplitude the metal substrate slightly fluctuates around its open circuit without significant perturbation of the surface conditions. Therefore, this technique can be considered as not-destructive. The impedance $Z(\omega)$ can be defined by the ratio between potential $V(\omega)$ and current signals $I(\omega)$:

$$Z = \frac{\Delta V}{\Delta I} = Z_0 \frac{V_0 \sin(\omega t)}{I_0 \sin(\omega t + \varphi)} = Z_{\text{real}} + iZ_{\text{img}} \quad (2)$$

For each frequency, a specific impedance Z can be defined. Analogous to electrical resistance, impedance is a measure of opposition of a system to a perturbation from its steady state. From the corrosion point of view, at each frequency the various electrochemical processes respond with different rates, enabling to distinguish them by a simply evaluation of impedance evolution during test. For this purpose, the impedance data can be represented in two ways:

- Nyquist spectrum: where a Z_{real} (pure resistive component) versus Z_{img} (capacitive and inductive components) plot is used.
- Bode spectrum: log $|Z|$ and phase angle ϕ versus log frequency.

As reference, a scheme of the Nyquist and Bode plot for a defective coating on a metal substrate under immersion in an electrolyte solution is reported in Fig. 3.

2.2.5 Electrochemical Noise

Electrochemical noise (ECN) is a non-destructive technique where current and potential fluctuations are recorded during free metal corrosion (without applying external signal excitation). A stochastic pulse of current or potential can be related

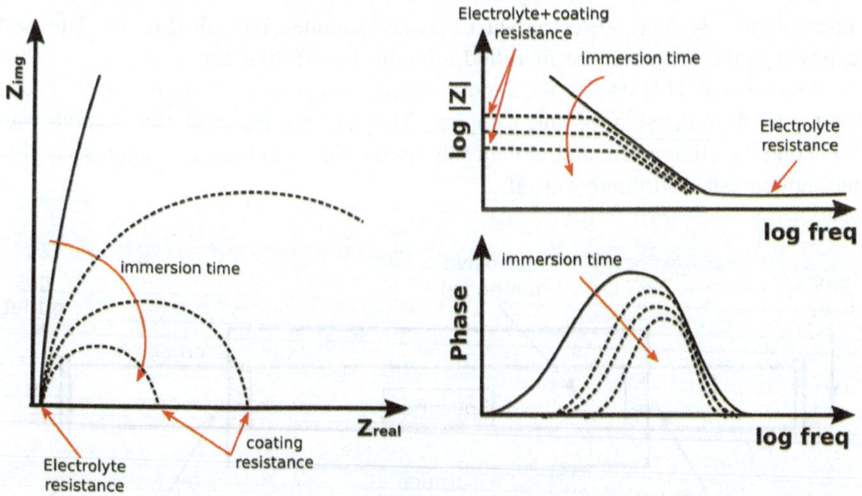

Fig. 3 Scheme of Nyquist (left) and Bode plot (right) for a defected coated metal substrate under immersion in an electrolyte solution

to a specific electrochemical corrosion phenomenon (metal dissolution, hydrogen evolution, pitting, etc.) [10]. To perform this technique the potential or current signal perturbations should be recorded between two nominally identical electrodes. Usually a three-electrode configuration is used. A two identical electrode configuration is also applicable. The current fluctuations are recorded between two identical metal electrodes. The potential fluctuations are recorded between one of the metal electrode and a standard reference electrode, that can be considered not signal perturbing.

3 Corrosion Characterization of Molten Salts Used as Materials Storage at High Temperature

Gravimetric studies need to follow the same protocol in order to obtain results that can be compared between them. In this direction, the International Standard ISO/NP 17245 [11] specifies the method for high-temperature corrosion testing of metallic materials by immersion in molten salt or other liquids which are under static conditions and exposing them to high temperature in a controlled gas environment to evaluate their corrosion resistance. Soleimani et al. [12] followed this standard for corrosion evaluation in molten salts. Figure 4 shows the corrosion setup and the coupons configuration into the salt.

In this case, the samples are placed at the bottom of the crucible and the exposed area should be corrected. In the standard, the sample must be placed in brackets (Fig. 5) but depending on the corrosive environment this fact is not trivial.

An innovative placement for these samples is focused on this type of immersion systems. Gurr et al. [13] held the samples with titanium wire and fixed them in an alumina tube, getting a direct contact between samples and salt (Fig. 6). The area exposed to the salt could be modified through this configuration.

Novello et al. [14] studied the vapour effect in a regular molten salt corrosion test with partial immersion of the samples. The authors reported that vapour salt (affecting the atmosphere and molten salt interphase) could be more aggressive that the molten salt environment itself.

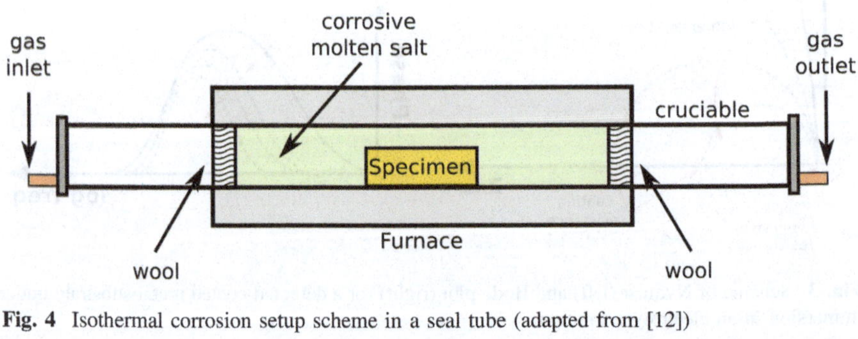

Fig. 4 Isothermal corrosion setup scheme in a seal tube (adapted from [12])

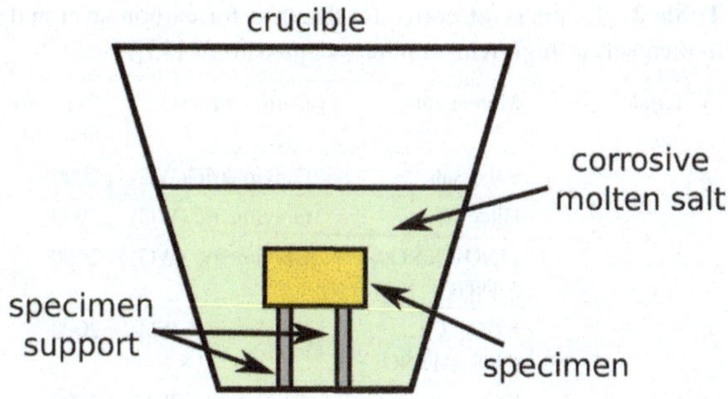

Fig. 5 Corrosion setup standard configuration (adapted from [11])

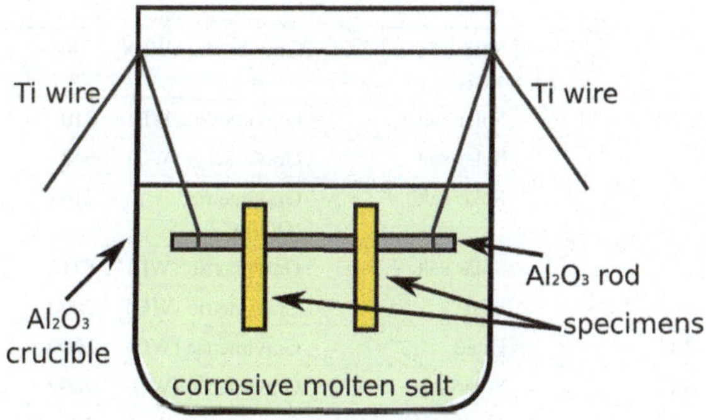

Fig. 6 Schematic representation of immersion in molten salt (adapted from [13])

An optimal material selection has to be performed in order to avoid cross contamination in the coupons tested; in addition, handling must be done carefully since this operation could remove the corrosion products obtained during the test.

Trying to solve this problem, a new methodology (Dynamic Gravimetric Analysis, DGA) has been developed at the University of Barcelona [7]. The sample is hung from an analytical balance till the cleaning solution starts to attack the metal after removing all the metal oxide. This point is detected by a slope change in the mass loss versus time registered obtaining a more accurate corrosion rate results than the ASTM G1-03 standard and avoiding technician handling.

A recent comprehensive review [15] unifies the corrosion rate obtained in the different studies present in the literature (Tables 2 and 3), including the corrosion elucidation techniques used, immersion time, test temperature and molten salt proposed.

Results shown in Tables 2 and 3 include carbon steels, stainless steels and alumina forming alloys. The elements diffusion as well as the protective or non-protective scale formed in the steel surface is usually analysed by scanning electron microscopy (SEM). Fernandez et al. [4] studied the multiscale form in conventional stainless steel (304) immersed during 2000 h in solar salt. Microstructure is shown in Fig. 7.

Table 2 Summary of corrosion data/test for carbon steel and stainless steel immersed in nitrate molten salt at high temperature (adapted from [15])

Material	Molten salt	Corrosion test	Exposure time (h)	Temperature (°C)	Corrosion rate (original unit)
A1	Solar salt	Gravimetric (WG)	2000	390	1.1 mg/cm^2
	Hitec XL	Gravimetric (WG)	2000	390	0.71 mg/cm^2
	$LiNO_3$–KNO_3–$NaNO_3$	Gravimetric (WG)	2000	390	1.125 mg/cm^2
	KNO_3–$Ca(NO_3)_2$–$LiNO_3$	Gravimetric (WG)	2000	390	0.85 mg/cm^2
A36	Solar salt	Gravimetric (WL)	7008	316	2.32 mg/cm^2
	Solar salt + 1.0 Cl wt%	Gravimetric (WL)	7008	316	3.14 mg/cm^2
	Solar salt + 1.3 Cl wt%	Gravimetric (WL)	7008	316	2.05 mg/cm^2
A516 Gr.70	Solar salt	Gravimetric (WL)	810	400	0.078 mm/year
	Solar salt	Gravimetric (WL)	642	450	0.299 mm/year
	Solar salt	Gravimetric (DGA)	2165	400	47.47 μm/year
	Solar salt	Gravimetric (WL)	8712	390	2.14 μm/year
	Hitec	Gravimetric (WG)	2000	390	0.31 mg/cm^2
T11	Hitec	Gravimetric (WG)	2000	390	0.43 mg/cm^2
T22	Solar salt	Gravimetric (WG)	2000	390	2.25 mg/cm^2
	Solar salt	Gravimetric (WG)	2000	550	Catastrophic after 800 h
	Hitec	Gravimetric (WG)	2000	390	0.31 mg/cm^2
	Hitec XL	Gravimetric (WG)	2000	390	0.21 mg/cm^2
	$LiNO_3$–KNO_3–$NaNO_3$	Gravimetric (WL)	1000	550	139 μm/year
	$LiNO_3$–KNO_3–$NaNO_3$	Gravimetric (WG)	2000	390	0.825 mg/cm^2
	KNO_3–$Ca(NO_3)_2$–$LiNO_3$	Gravimetric (WG)	2000	390	0.525 mg/cm^2
T5	$LiNO_3$–KNO_3–$NaNO_3$	Gravimetric (WL)	1000	550	110 μm/year
T9	$LiNO_3$–KNO_3–$NaNO_3$	Gravimetric (WL)	1000	550	13.3 μm/year
X20CrMoV12-1	$LiNO_3$–KNO_3–$NaNO_3$	Gravimetric (WL)	1000	550	5.5 μm/year
X20CrMoV11-1	Solar salt	Gravimetric (WG)	2500	600	18.1 mg/cm^2
P91	Solar salt	Gravimetric (WG)	2500	600	21.1 mg/cm^2
P92	Solar salt	Gravimetric (WG)	500	600	0.85%

For a proper corrosion study, corrosion compounds must be confirmed by XRD, usually performed in the steel surface and in the salt remaining as well. Compounds found in the salt remaining after the test could give information about the protective layer formed and their resistance over time.

4 Corrosion Characterization in TES Material at Low/Medium Temperature

4.1 Phase Change Materials

4.1.1 Immersion Tests

This section has the objective of summarizing the immersion test techniques used so far in the literature.

In this chapter, not only metals and metal alloys but also plastic containers [16] and its corrosion issues are implemented. In addition, other PCM nature; paraffin wax [17], fatty acids [18–20], molecular organic alloys [16], salt hydrates [16, 19, 21–27] are also considered.

The PCMs studied have a temperature range between 10 °C [26] and 60 °C [20], usually for cooling and heating applications respectively.

The different steps of the methodology used by most authors and summarized here are based in the same nomenclature and distribution as Krishna et al. [17]. The experimental methodology is divided in four steps: (1) the sample preparation, (2) corrosion test of the hypothetic container with the PCM (3) sample cleaning-up, and (4) assessment of the corrosion rate:

(1) Sample preparation.

- The samples are cut, polished [19, 27] and measured.
- They are cleaned using a cloth [27], distilled water [24] or acetone [18, 19, 24, 26] to remove dust, grease and surface dirt or the oil from the cutting process.
- It is desirable to have large surface-to-mass ratio and small ratio of the edge area to total area. Sharp edges and burrs of the faces of the samples can be avoided by grinding with abrasive paper [17].

(2) Corrosion test (container-PCM)

The immersion corrosion test serves the purpose of determining corrosion rates for low temperature application.

Lázaro et al. [16] studied plastic containers. The methodology is quite different. The PCM was melted and stirred in order to have a homogeneous sample. Then, a certain volume sample was introduced into each different plastic bottle.

Table 3 Summary of corrosion data/test for alloys immersed in nitrate molten salt at high temperature [36]

Alloy	Molten salt	Corrosion test	Exposure time (h)	Temperature (°C)	Corrosion rate
304	Solar salt	Gravimetric (WL)	7008	570	4.86 mg/cm^2
	Solar salt + 1.0 Cl wt%	Gravimetric (WL)	7008	570	7.31 mg/cm^2
	Solar salt + 1.3 Cl wt%	Gravimetric (WL)	7008	570	9.76 mg/cm^2
	Solar salt	Gravimetric (WL)—thermal cycling test	4432	565	0.00178 μm/h
	Solar salt + 0.82 Cl wt%	Gravimetric (WL)—thermal cycling test	4432	565	0.00372 μm/h
	Solar salt	Gravimetric (WL)—isothermal test	4584	565	0.0013 μm/h
	Solar salt + 0.82 Cl wt%	Gravimetric (WL)—isothermal test	4584	565	0.00358 μm/h
	Solar salt	Gravimetric (WG)	2000	390	0.01 mg/cm^2
	Solar salt	Gravimetric (WG)	2000	550	0.72 mg/cm^2
	Solar salt	Gravimetric (WL)	7008	570	4.86 mg/cm^2
	Solar salt	Gravimetric (WG)	1700	600	0.12 mg/cm^2
	Hitec XL	Gravimetric (WG)	2000	390	0.014 mg/cm^2
304 HCu	Solar salt	Gravimetric (WG)	500	600	0.19%
310 N	Solar salt	Gravimetric (WG)	500	600	0.19%
316	Solar salt	Gravimetric (WL)	7008	570	5.85 mg/cm^2
	Solar salt + 1.0 Cl wt%	Gravimetric (WL)	7008	570	5.15 mg/cm^2
	Solar salt + 1.3 Cl wt%	Gravimetric (WL)	7008	570	5.56 mg/cm^2
	Solar salt	Gravimetric (WL), TC	4084	565	0.001224 μm/h
	Solar salt + 0.82 Cl wt%	Gravimetric (WL), TC	4084	565	0.00257 μm/h
	Solar salt	Gravimetric (WL)	4584	565	0.001025 μm/h
	Solar salt + 0.82 Cl wt%	Gravimetric (WL)	4584	565	0.001592 μm/h
	Solar salt	Gravimetric (WL)	3000	565	0.44 wt%
	Solar salt + 0.25 Cl wt%	Gravimetric (WL)	3000	565	1.05 wt%
	Solar salt	Gravimetric (WL)	3000	600	0,00143 μm/h
	Solar salt	Gravimetric (WG)	5000	600	0.55 mg/cm^2

(continued)

Table 3 (continued)

Alloy	Molten salt	Corrosion test	Exposure time (h)	Temperature (°C)	Corrosion rate
316L	Solar salt	Gravimetric (WL), TC	4084	565	0.00139 μm/h
	Solar salt + 0.82 Cl wt%	Gravimetric (WL), TC	4084	565	0.0014 μm/h
	Solar salt	Gravimetric (WL)	4584	565	0.000959 μm/h
	Solar salt + 0.82 Cl wt%	Gravimetric (WL)	4584	565	0.0109 μm/h
321	Solar salt	Gravimetric (WG)	3000	400	1 μm/year
	Solar salt	Gravimetric (WG)	3000	500	7.1 μm/year
	Solar salt	Gravimetric (WG)	3000	600	15.9 μm/year
	Solar salt	Gravimetric (WG)	3000	680	460 μm/year
	Solar salt	Gravimetric (WL)	3000	565	0.23%
	Solar salt + 0.25 Cl wt%	Gravimetric (WL)	3000	565	0.75%
	Hitec	Gravimetric (WL)	2000	530	0.018 mm/year
347	Solar salt	Gravimetric (WG)	3000	400	0.7 μm/year
	Solar salt	Gravimetric (WG)	3000	500	4.6 μm/year
	Solar salt	Gravimetric (WG)	3000	600	10.4 μm/year
	Solar salt	Gravimetric (WG)	3000	680	447 μm/year
347H	Solar salt	Gravimetric (WL)	3000	600	0.00158 μm/h
	Solar salt	Gravimetric (WG)	5000	600	0.46 mg/cm^2
OC-4	Solar salt	Gravimetric (WG)	2000	390	0.22 mg/cm^2
	Solar salt	Electrochemical (EIS)	340	390	10.54 Ω

This step includes:

- The samples can be fully or partially [24] immersed in the PCM, with different purpose studies. The presence of oxygen in the uppermost zone of the specimens, in comparison with the low oxygen content available in the molten PCM can act as a (galvanic) concentration cell. The conclusion of a more corrosive effect on the container by volatile than solid matter of the PCM is not clear. A weight loss gravimetric analysis must be carried out to confirm this statement since salts produce spallation in the corrosion layer usually formed, so more presence of this layer could still be visible in the uppermost zone.
- In order to conduct the experiment in a liquid phase condition of the PCM the test can be either heating over the melting point and/or putting the samples through thermal and phase change cycles [16, 18, 20]. With that purpose thermal baths [17, 18, 21–23, 28] or stove and furnaces [16, 19, 20, 26] were used.

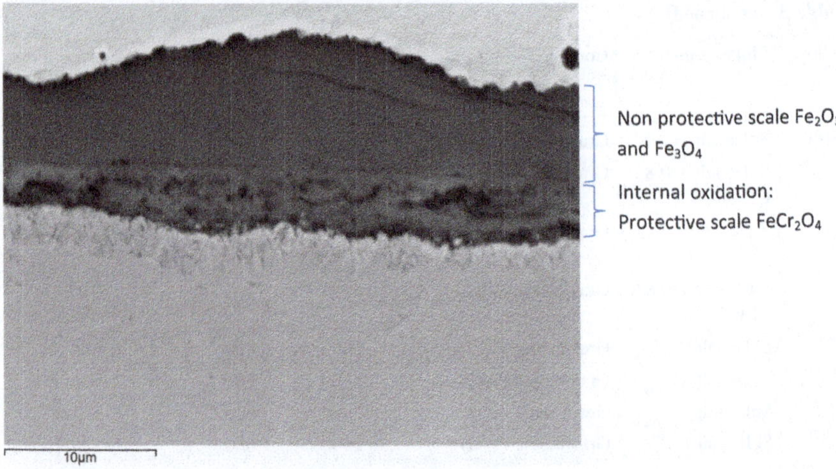

Fig. 7 Cross sectional microstructure obtained by SEM in 304SS immersed in solar salt (60% $NaNO_3$ + 40% KNO_3) after 2000 h (adapted from [4])

- There is no fixed criterion on the temperature set for the corrosion characterization experiments. Cabeza et al. [21, 22, 28] proposed a temperature 20 °C above the PCM melting point [21]. The temperature selected for Garcia-Romero et al. is the maximum temperature calculated to be reached in peak periods in the designed system application [24]. All papers studied and the book authors agree to the necessity of a test temperature over the PCM tested melting point to have it in liquid conditions, the range is among 13 °C [24] and 55 °C [23] temperature increment.
- The duration of the experiment is not fixed either.

The longest tests would be Lázaro et al. [16] or Browne et al. [18] which perform a two years or ten months tests which implies 103 or 150 phase change cycles, respectively. Also Cabeza et al. [23] carried out a series of one and a half years test in a liquid state temperature range of the PCM.

In order to extrapolate experimental values for a year time period [25], at least a 2000 h corrosion test characterization must be performed. The curve of the corrosion rate over time is not a linear one; it usually follows a parabolic behaviour. Therefore, a first stabilization time period is required. This means, 3 months representative results, no less, could be extrapolated to one year time period, and still the nature of the specific PCM-container corrosion kinetics should be studied before pulling off positive results.

(3) Sample cleaning-up and evaluation

For each cycle [16, 18, 20, 27] or at the end of the corrosion test each researcher used different procedures as for the preparation of the sample step. Cabeza et al. [21–23] and Moreno et al. [26] proposed several evaluation procedures dealing with salt hydrate PCMs and metallic containers. Visual evaluation seeking for bubbles, precipitates, surface changes, pitting processes, and a pH measure of the PCM solution are recommended. After that, a thoroughly cleaning using tap water and abrasive paper polish if necessary are conducted, Ferrer et al. [19] recommend specimens to be dried with soft paper previous a detailed optical observation or a corrosion rate assessment.

Most authors used the guidelines of ASTM G1 standards [5] for preparing, cleaning, and evaluating corrosion test specimens, G1-03 [18, 19, 24, 26]. No international standards on PCM migration in plastics tests were found so the standard ISO 175:1999 Plastics, Methods of Test Determination of the Effects of Immersion in Liquid Chemicals was used in Lázaro et al. [16].

(4) Assessment of the corrosion rate

As stated in Sari et al. [20] gravimetric analysis, used for all existing investigation, after the corrosion tests provide the mass loss from which the reduction in thickness with time (for uniform surface corrosion) and also the obtained data is used to calculate corrosion rate, taking onto account the material density in contact with the corrosive environment.

Microscope technique, optic [26, 27], metallographic [20] or stereoscopic [24] are used, on the other hand, to provide information on the type of corrosion (e.g. pitting corrosion, cracks, etc.).

A microstructural analysis in scanning electron microscope SEM can also be found for a specific surface characterization in [17, 18].

Tables 2, 3, 4, 5, 6 and 7 summarize all results on corrosion test compatibility between several containers and PCM combinations found in the literature.

4.1.2 Electrochemical Tests

Experimental Setup

For the electrochemical testing of PCM materials a standard approach can be identified in a three-electrode system experimental setup implemented with a frequency analysis equipped potentiostat instrument (in these tests a Biologic SP300 potentiostat equipped with low current and EIS module was used).

The investigated metal (working electrode) is directly connected to the potentiostat by a clamp. The counter electrode can be a stable non-reactive material as platinum wires. As reference electrode a conventional standard reference electrode (e.g. Ag/AgCl electrode) can be used. However, in order to avoid electrode contamination during the measurement a glass protected platinum wire can be used.

Table 4 Summary of the results of corrosion between PCM and container—salt hydrate/metal

PCM	Melting temperature (°C)	Copper	Brass	Aluminium	Carbon steel	Stainless steel-316	Stainless steel-304
TH29 ($CaCl_2{\cdot}2H_2O$)	n.a.	Recommended [23]	Recommended [23]	Not recommended [23]	Not recommended [23]	n.a.	Recommended (caution, possible precipitate) [23]
TH29 ($CaCl_2{\cdot}2H_2O$) + $MgCl_2{\cdot}6H_2O$	n.a.	Recommended [23]	Recommended [23]	Not recommended [23]	Not recommended [23]	n.a.	Recommended (caution, possible precipitate) [23]
S10 (Na_2SO_4 + NH_4Cl + sepiolite)	10	Not recommended [26]	n.a.	Caution recommended [26]	Not recommended [26]	Recommended [26]	n.a.
C10 (Na_2SO_4 + H_2O + additives)	10.5	Not recommended [26]	n.a.	Recommended [26]	Not recommended-long term [26]	Recommended [26]	n.a.
$ZnCl_2{\cdot}3H_2O$	10	Recommended [26]	n.a.	Not recommended [26]	Not recommended-long term [26]	Recommended [26]	n.a.
$K_2HPO_4{\cdot}6H_2O$	13	Caution recommended [26]	n.a.	Not recommended [26]	Not recommended [26]	Recommended [26]	n.a.
$NaOH{\cdot}1.5H_2O$	15	Not recommended [26]	n.a.	Not recommended [26]	Caution recommended [26]	Recommended [26]	n.a.
$K_3PO_4{\cdot}7H_2O$	45	Not recommended [26]	n.a.	Not recommended [26]	Recommended [26]	Recommended [26]	n.a.
$Zn(NO_3){\cdot}4H_2O$	45.5	Not recommended [26]	n.a.	Not recommended [26]	Not recommended [26]	Recommended [26]	n.a.
S46 ($Na_2S_2O_3{\cdot}5H_2O$ + sepiolite + fumed silica)	46	Not recommended [26]	n.a.	Caution recommended [26]	Not recommended-long term [26]	Recommended [26]	n.a.
$Na_2S_2O_3{\cdot}5H_2O$	48	Not recommended [22, 26]	Not recommended [22]	Caution recommended [22, 26]	Recommended [22] Caution recommended [26]	Recommended [26]	Recommended [22]

(continued)

Table 4 (continued)

PCM	Melting temperature (°C)	Copper	Brass	Aluminium	Carbon steel	Stainless steel-316	Stainless steel-304
C48 (CH_3OONa + H_2O + additives)	48	Not recommended [26]	n.a.	Recommended [26]	Recommended [26]	Recommended [26]	n.a.
$MgSO_4{\cdot}7H_2O$	48.5	Not recommended [26]	n.a.	Recommended [26]	Not recommended [26]	Recommended [26]	n.a.
C18 ClimSel (CH_3COONa + additives)	18	The copper samples experienced the greatest mass loss [27]	n.a.	The corrosion rate in the C18 immersed aluminium alloy samples increased by, on average, a factor of 8 [27]	n.a.	n.a.	n.a.
E17 PlusICE ($Na_2SO_4{\cdot}10H_2O$ + NaCl)	17	The copper samples experienced the greatest mass loss [27]	n.a.	Negligible mass was lost from the aluminium samples in contact with E17 although localised pitting with this PCM was observed from the metallographic examination [27]	n.a.	n.a.	n.a.
CH_3OONa	58	Caution recommended [22]	Caution recommended [22]	Recommended [22]	Recommended [22]	n.a.	Recommended [22]
$Zn(NO_3)_2{\cdot}6H_2O$	36	Not recommended [21]	Not recommended [21]	Not recommended [21]	Not recommended [21]	n.a.	Recommended [21]
$Na_2HPO_4{\cdot}12H_2O$	35	Caution recommended [21]	Recommended [21]	Not recommended [21]	Caution recommended [21]	n.a.	Recommended [21]
$CaCl_2{\cdot}6H_2O$	32	Recommended [21]	Recommended [21]	Caution recommended [21]	Caution recommended [21]	n.a.	Recommended [21]

Table 5 Summary of the results of corrosion between PCM and container—fatty acid/metal

PCM	Melting temperature (° C)	Cooper	Brass	Aluminium	Mild steel	Stainless steel-316	Stainless steel-304L	Carbon steel-C20	Perspex
Stearic acid	53.8	Slightly corroded [20]	n.a.	Resistant [20]	n.a.	n.a.	Resistant [20]	Resistant [20]	n.a.
Palmitic acid	59.9	Slightly corroded [20]	n.a.	Resistant [20]	n.a.	n.a.	Resistant [20]	Slightly corroded [20]	n.a.
Mystiric acid	53.8	Slightly corroded [20]	n.a.	Resistant [20]	n.a.	n.a.	Resistant [20]	Slightly corroded [20]	n.a.
Lauric acid	42.6	Slightly corroded [20]	n.a.	Resistant [20]	n.a.	n.a.	Resistant [20]	Resistant [20]	n.a.
Capric acid	27.5–32.7	Caution [18]	Caution [18]	Caution [18]	Caution [18]	Recommended [18]	n.a.	n.a.	Recommended [18]
Capric and palmitic acid	17.7–22.8	Caution [18]	Caution [18]	Caution [18]	Caution [18]	Recommended [18]	n.a.	n.a.	Recommended [18]
Capric and lauric acid	18.9–22.3	Caution [18]	Caution [18]	Caution [18]	Caution [18]	Recommended [18]	n.a.	n.a.	Recommended [18]

Table 6 Summary of the results of corrosion between PCM and container—other organic/metal

PCM	Melting temperature (°C)	Copper	Brass	Aluminium	Mild Steel	Stainless steel-316	Perspex
SP22 eutectic paraffin: fatty acid	16.9–25.3	Recommended [18]	Recommended [18]	Not recommended [18]	Recommended [18]	Recommended [18]	Recommended [18]
Micronal	20.9–25.8	Recommended [18]	Recommended [18]	Recommended [18]	Caution [18]	Recommended [18]	Recommended [18]
Paraffin wax	–	Recommended [17]	n.a.	n.a.	n.a.	n.a.	n.a.

Table 7 Summary of the results of corrosion between PCM and container—organic/plastic

PCM	Melting temperature (°C)	LDPE	HDPE	PET	PP
Molecular alloy 34% C16–66% C18	19.5–22.2	Not recommended [16]	n.a.	Recommended [16]	n.a.
RT20	20	Not recommended [16]	n.a.	Recommended [16]	Caution [16]
RT26	26	Not recommended [16]	n.a.	Recommended [16]	Caution [16]
RT25	25	Not recommended [16]	n.a.	Recommended [16]	Caution [16]
(TH24 inorganic salt)	24	Not recommended [16]	Recommended [16]	Not recommended [16]	Not recommended [16]

The heating of the sample is achieved by a laboratory hotplate. By a temperature controller it has been possible to maintain the temperature constant. A large working electrode surface area is required to avoid heterogeneous electrochemical activity induced by crevice phenomena or not ideal wetting by molten salt with the metal substrate. For the proposed experimental campaign, the exposed area, for all the samples, was 16 cm^2. After mechanical grinding with emery paper at 800 grid a chemical surface pre-treatment were performed for each sample as in the following:

- Aluminium alloy—Al (AA6061) samples were immersed for 1 min at room temperature in 0.1 M NaOH solution;
- Copper alloy—Cu (CW024A) samples were immersed for 15 min in a 25 wt% HNO_3 solution.
- Carbon Steel—CS (AISI 1050) samples were immersed for 20 min in trichloroethylene (sonication) and after for 15 min with a 15 wt% solution of HNO_3.

Details of surface pre-treatments are summarized in Table 8.

As reference PCM material was chosen the PlusIce S83 PCM molten salt (magnesium nitrate hexahydrate). The following electrochemical tests were carried out at increasing immersion time up to 7 days at 120 °C:

- Electrochemical impedance measurements (EIS) were carried out at OCP with a voltage amplitude of 10 mV and a frequency range from 0.05 to 100 kHz (10-point for each frequency decade were acquired).
- OCP measurement versus immersion time.

A scheme of the experimental setup is reported in Fig. 8.

Table 8 Surface pre-treatments steps for all metal alloys investigated

AA 6061	CW024A	AISI 1050
• 1 min@25 °C in 0.1 M NaOH solution • Washed in water • Dried with ethanol	• 15 min@25 °C in 25wt% HNO_3 solution • Washed in water • Dried with ethanol	• 20 min@25 °C in trichloroethylene under sonication • Washed in ethanol • 15 min@25 °C in 25wt% HNO3 solution • Washed in water • Dried with ethanol

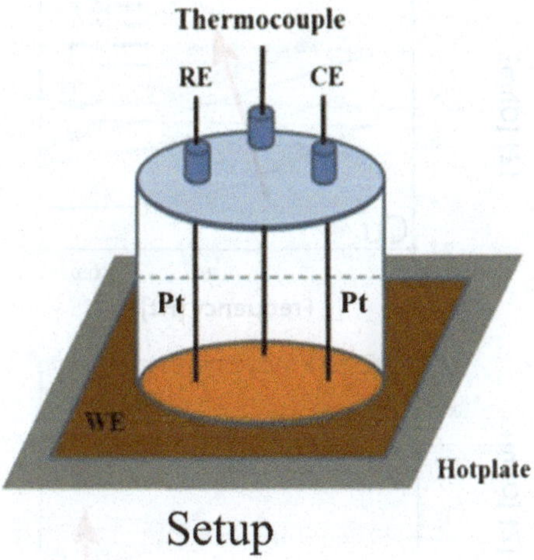

Fig. 8 Scheme of the electrochemical cell setup for metal corrosion tests in molten salt

At the end of the test, all samples were cleaned with distilled water and acetone and finally stored in a plastic bag with silica gel desiccant in order to preserve surface until performing the selected surface analysis (morphological analysis by optical, electron microscopy, XRD, FTIR analysis, etc.). Eventually, to determine the extent of pitting, to predict the remaining life of the metal component, or to select the most pitting-resistant materials for service the cleaning procedure suggested by ASTM G1 standard [5] can be carried out.

Results of the Case Study

In Fig. 9 the plot of the impedance modulus plot at varying frequency (Bode modulus plot) combined to OCP trend at increasing immersion time in the molten salt at 120 °C are shown.

For the aluminium sample, by EIS test, for the aluminium sample, it was possible to observe a slight but steady increase in the impedance modulus values for high and low frequency at the increasing of the immersion time in molten salt. The analysis of the potential versus time (EVT) confirms a small increase (less of

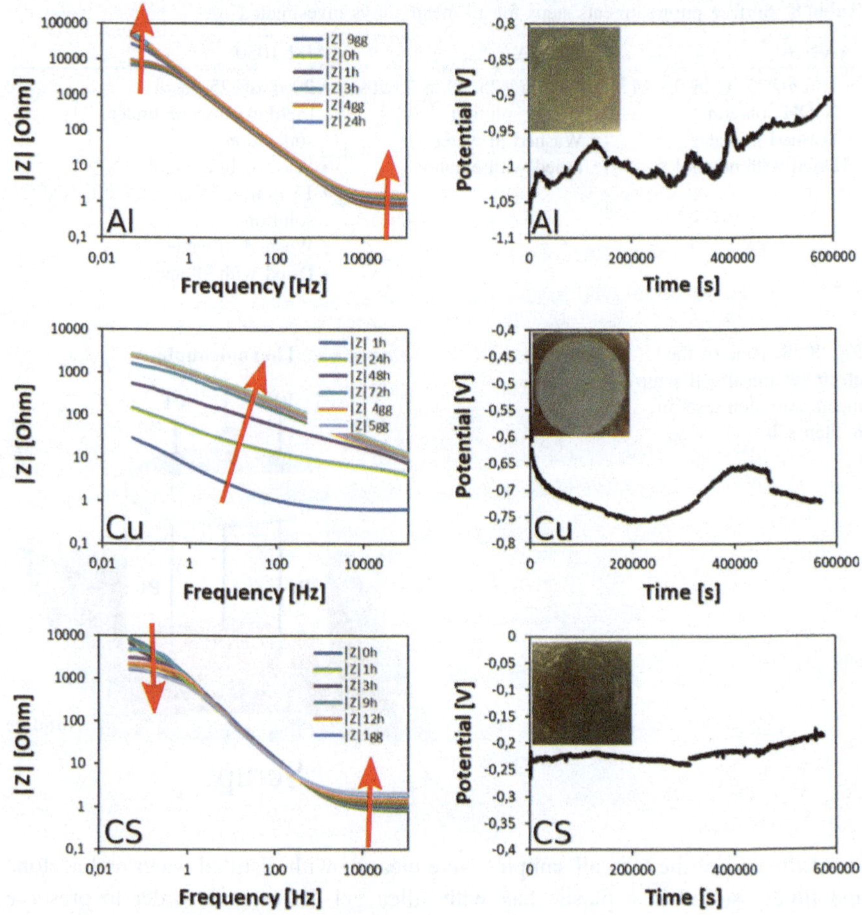

Fig. 9 Bode Impedance modulus plot and OCP trend plot for aluminium, copper and carbon steel samples

80 mV) of the potential at increasing time probably due to the formation, on the metal surface, of a thin protective oxide layer. The morphological analysis, performed by a stereomicroscope, showed a non-defective surface. No pits were identified.

The EIS of copper shows a rapid increasing of the |Z| for all the frequency range (0.05 Hz–100 kHz) during the first 72 h of immersion (≈2 decades). After this period, there was a stabilization of the behaviour. The potential trend has an initial decrease until 60 h and then an increase until a value of about −650 mV. By X-ray diffraction analysis of the copper sample surface (Fig. 10) it was possible to evidence the presence some copper compounds (diffraction peaks from copper substrate are also present).

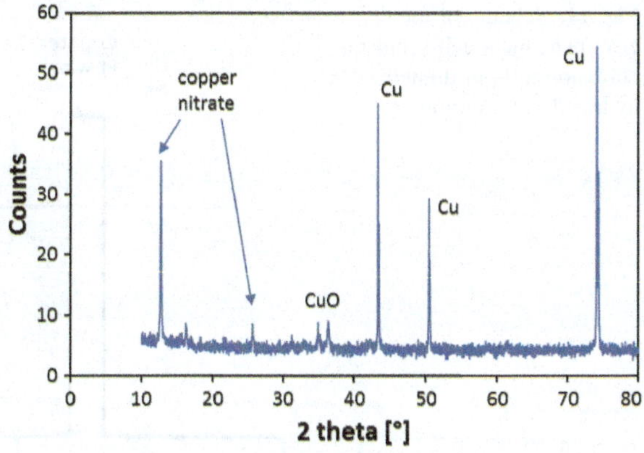

Fig. 10 X-ray diffraction patter of corroded copper surface

The carbon steel sample shows a little increase of impedance at high frequency and a decrease at low frequency (<1 decade). At increasing immersion time, the potential remains almost constant. The low oxide formation on the metal surface and the quite low weight loss confirm his acceptable performance in this media.

4.2 *Sorption and Chemical Reaction Materials*

4.2.1 Experimental Setup

In the following a possible test configurations to assess electrochemical and corrosion stability on different types of metal substrates in hydrated salts solutions has been reported.

The electrochemical tests, by using a three conventional electrode setup, need to be carried out by using an frequency analysis equipped potentiostat instrument (in these tests a Biologic SP300 potentiostat equipped with low current and EIS module was used).

- Working electrode: metal substrate under investigation.
- Reference electrode: Ag/AgCl sat.
- Counter Electrode: Pt wire.

In order to have a liquid electrolyte to carry out electrochemical testing diluted salt solutions were used instead of using pure salts. The electrolyte used was 10% $MgSO_4$ hexahydrate or $SrBr_2$ heptahydrate solution alternatively. A scheme of the electrochemical cell is reported in Fig. 11.

Considering the high conductivity of the electrolyte solution a small exposed surface of the working electrode can be used (a surface of about 0.75 cm^2 was selected in this case). To have high repeatable results a periodic regeneration of the solution during time is required, in order to have a quite constant oxygen content during the test.

Fig. 11 Scheme of the electrochemical cell setup for corrosion tests in diluted hydrated salt solution

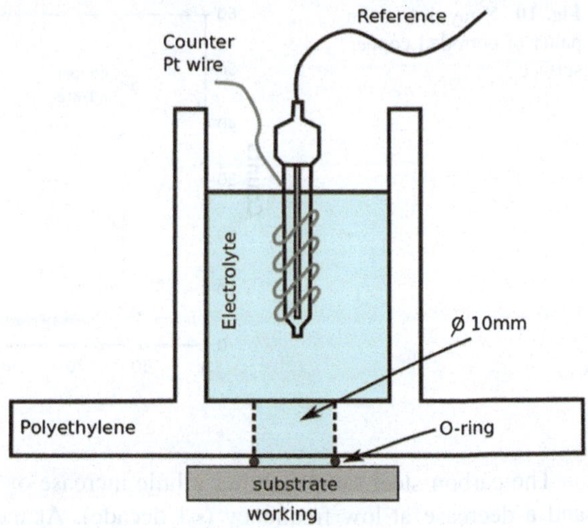

In this configuration, a minimum dimension of each sample should be 25 mm × 25 mm. No specific constrain on thickness are required. The water based electrolyte solution chosen allows applying a wider spectrum of electrochemical analysis methods than that carried out in molten salts case study.

In this case, study, for each sample, the electrochemical characterization was configured in accordance to a six steps procedure reported in Table 9.

According to ASTM G61 [29], a 1 h stabilization procedure was proposed. During this time the open circuit potential (OCP) was monitored in order to verify the constancy of the electrochemical potential of the sample and to identify possible instability condition that prevent the application of the following steps. Then an electrochemical impedance spectroscopy (EIS) test will be performed in the frequency range 50 mHz–100 kHz to study the surface properties of the substrate.

Afterwards, the potentiodynamic cyclic test in anodic direction was performed. In particular the polarization starts at −100 mV respect to OCP, moving toward the anodic potentials at a speed of 0.2 mV/s. After reaching a specific current value (a threshold value 0.1 mA/cm^2 was here used), the polarization will be reversed to cathodic direction to evaluate the material's ability to repassivate. By this approach, the sensitivity of the substrates to pitting and repassivation in the electrolytes can be evaluated.

Finally, the stabilization period (1 h with EVT control) and EIS test are carried out again in order to evaluate the electrochemical properties of the 'repassivate' substrate. All tests were performed at room temperature and in open to air condition. Of course, a different rearrangement of the testing procedure proposed in Table 9 can be carried out based on specific research aims, e.g. a possible improvement can include the performance of LPR tests in order to evaluate corrosion rate of the working electrode during immersion time. In the present case

Table 9 Electrochemical testing procedure applied for corrosion performance evaluation of investigated metal alloys

Sample conditioning	Test code	Notes
EIS	EIS	Electrochemical impedance measurement in the frequency range 50 MHz–100 kHz
1 h stabilization at free potential	EVT	OCP will be calculated
EIS	EIS	Electrochemical impedance measurement in the frequency range 50 MHz–100 kHz
Test according to ASTM G61	Potentiodynamic	Anodic branch. E_{pit} and E_{pro} will be calculated
1 h stabilization at free potential	EVT	OCP will be calculated
EIS	EIS	Electrochemical impedance measurement in the frequency range 50 MHz–100 kHz

Table 10 Electrolyte solutions used in LPR tests

Testing environment	Salt content (wt%)	Testing temperature (°C)
$SrBr_2.6H_2O$	10	25
$MgSO_4.7H_2O$	10	25

study, the tests will be performed by using the two electrolyte solutions listed in Table 10.

After mechanical grinding with emery paper at 800 grid, a chemical surface pre-treatment were performed for each metal alloy substrate investigated:

- Aluminium alloy—Al (AA6061) samples were immersed for 1 min at room temperature in 0.1 M NaOH solution, then washed with distilled water and dried with ethanol.
- Copper alloy—Cu (CW024A) samples were immersed for 15 min in a 25 wt% HNO_3 solution, then washed with distilled water and dried with ethanol.
- Carbon Steel—CS (AISI 1050) samples were immersed for 20 min in trichloroethylene (sonication) and after for 15 min with a 15 wt% solution of HNO_3, then washed with distilled water and dried with ethanol.
- Stainless Steel—SS (AISI 304) samples were immersed for 20 min in trichloroethylene (sonication) and after for 15 min with a 15 wt% solution of HNO_3, then washed with distilled water and dried with ethanol.

At the end of the electrochemical testing procedure, reported in Table 9 all samples were cleaned with distilled water and acetone and finally stored in a plastic bag with silica gel desiccant in order to preserve surface until performing the selected surface analysis (morphological analysis by optical, electron microscopy, XRD, FTIR analysis, etc.).

Results of the Case Study

Preliminary information concerning electrochemical behaviour of a metal in hydrated salt solution can be acquired analysing the electrochemical impedance spectroscopy trend at increasing immersion time. In Fig. 12a the comparison of EIS curve (impedance modulus in Bode plot) after 0 and 1 h of immersion in magnesium sulphate solution is reported for carbon steel. The magnitude of the modulus at low frequency is related to the polarization resistance of the metal substrate. In particular, the curve obtained at 0 h evidenced a maximum magnitude of impedance modulus of about 1000 Ω at 1 Hz. At lower frequency, a progressive reduction of its value can be observed. This behaviour can be related to a not stationary behaviour of the surface. The surface evidences the presence of the so called inductive contribute due to the metal dissolution in the electrolyte solution and the subsequent formation of corrosion products that precipitate on the surface itself modifying its electrochemical activity [30]. After 1 h of immersion, the corrosion products become significant thus reducing the mass transport phenomena. The impedance modulus increases of about one order of magnitude, and a mainly capacitive behaviour can be observed (monotone growing linear trend in the modulus Bode plot). Furthermore, in Fig. 12b cyclic polarization curve is shown.

The corrosion potential, OCP, is about −0.700 V versus Ag/AgCl sat. The corrosion current esteemed by Tafel approach was about 2×10^{-3} mA/cm^2. The ascending and descending polarization curve did not evidence any particular modifications indicating that a uniform corrosion was occurring on the metal surface.

In order to compare the electrochemical behaviour of each metal alloy in the salt based electrolyte solution a comparison plot of EIS and polarization curves is reported in Fig. 13a, b, respectively. Analysing the impedance modulus trend, the worst corrosion resistance was observed for carbon steel substrate. The most suitable result was obtained for stainless steel and aluminium alloy. Furthermore,

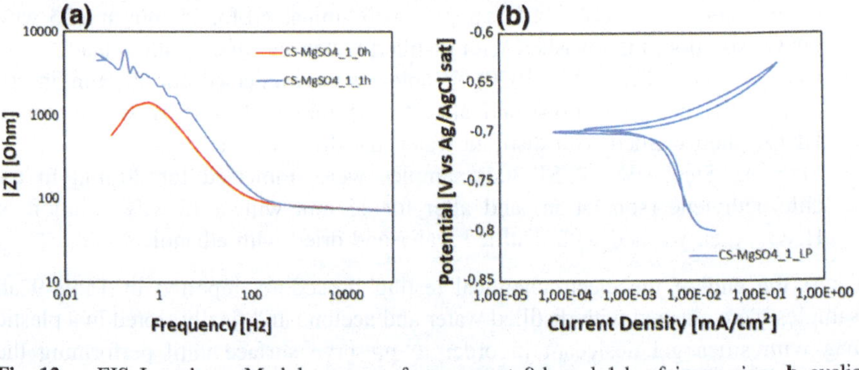

Fig. 12 **a** EIS Impedance Modulus versus frequency at 0 h and 1 h of immersion; **b** cyclic polarization curve after 1 h of immersion. Carbon Steel substrate in 10% magnesium sulphate solution

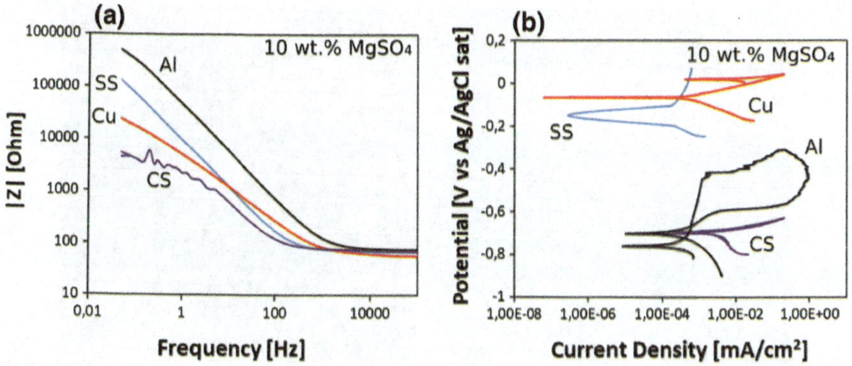

Fig. 13 EIS Impedance **a** Modulus versus frequency, and **b** Ascending and descending polarization; after 1 h of immersion in 10% magnesium sulphate solution for all metal substrates

evaluating the polarization curves obtained for all metal substrate is evident that materials are clustered in two classes. Carbon steel (CS) and aluminium alloys (Al) identify the low noble materials, characterized by a low corrosion potential. Instead, stainless steel (SS) and copper (Cu) alloys identify the more noble materials, characterized by higher corrosion potential.

Furthermore, low corrosion current (and consequently low corrosion rate) was observed for stainless steel sample, where i_{corr} of about 1×10^{-4} mA/cm^2 was calculated. Large corrosion rate was observed for carbon steel and copper alloys. These considerations are confirmed by stereo images of corroded surfaces (Fig. 14). Carbon steel and copper alloy substrates shows evident corrosion products deposited homogeneously overall on the surface. Conversely, aluminium alloys show some pits indicating metal sensitivity to localized corrosion attack. Finally, stainless steel substrate did not evidence any attack suggesting a quite stable electrochemical behaviour in this electrolyte solution.

Similar consideration on corrosion metal stability can be acquired for 10 wt% strontium bromide solution for all metal substrates. In particular, carbon steel alloy resulted sensible to corrosion attack in both solution, evidencing a higher electrochemical reactivity in $MgSO_4$-based solutions than in $SrBr_2$ based one.

The aluminium alloy has a higher electrochemical sensitivity to the Strontium bromide electrolyte solution. The halide ions contribute to decrease the polarization resistance thus limiting alloys stability in the solution [31]. These considerations can be evidenced also by a slight reduction of impedance modulus at low frequency. Conversely, the noble metal substrates (Cu and SS) did not show any specific sensitivity to halide solutions. In particular, stainless steel metal showed a very good electrochemical stability in both solutions confirming the very good corrosion resistance of this substrate in aggressive salt hydrated solutions. In both cases on copper substrates a thick oxide layer formed due to higher electrochemical reactivity of the metal, but in this case oxide film positively influenced the

Fig. 14 Stereo-image of corroded surface area after corrosion tests in 10% magnesium sulphate solution for all metal substrates

electrochemical processes at the copper surface hindering mass transport phenomena with subsequent delay in corrosion degradation [32].

5 Corrosion Mitigation

Corrosion mitigation is one of the most important issues to control in the new generation of TES materials for CSP plants. Different corrosion protection methods are available in order to reduce the corrosive potential. The different protection possibilities are shown in Fig. 15.

A proper system design avoiding curves and heterogeneous zones is important to reduce molten salt corrosion as well as a good material selection. In addition, experimental conditions should be taken into account (temperature, pressure, time of exposure, etc.).

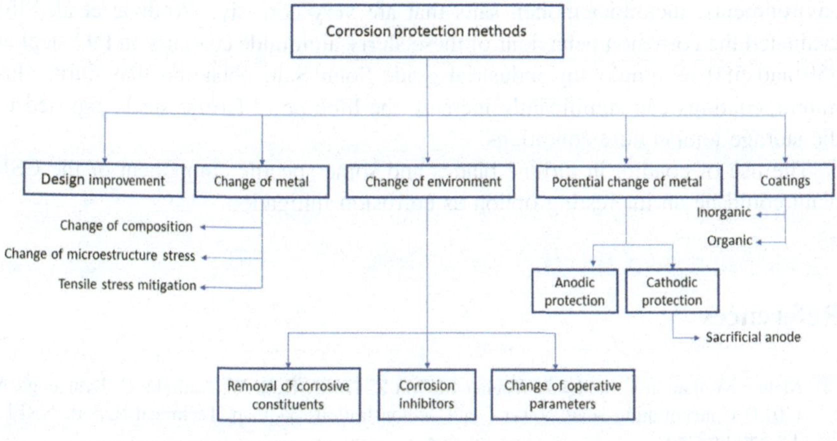

Fig. 15 Corrosion protection methods

In this direction, alumina forming alloys are currently being evaluated as an alternative to mitigate corrosion in molten salts at high temperatures (650–750 °C). The principle works through the formation of a native alumina scale that offers oxidation resistance superior to that of chromia-forming of conventional stainless steels. Magnesium (Mg) is able to control corrosion of alloys at the grain boundaries. Corrosion inhibitors such as Mg maintain corrosion at low levels because their oxidation is more thermodynamically favoured than this of the alloy components, such as Chromium, which is also highly desirable.

The use of corrosion inhibitors (CIs) added to the salt has been extended in the last years along with the use of MgO, CeO_2 and DyO_2 to help reduce corrosion in the alloys [33]. A novel approach in this direction could be included regarding the salt redox potential control. Some researchers have highlighted a new way for corrosion mitigation by adding different redox couples to fix the salt potential [34], this implementations could be the key for future developments since it implies the variation in the molten salt corrosive potential.

Another important research line related with corrosion mitigation in CSP plants using molten salts is the use of coating. Coatings could be a suitable solution in order to reduce the corrosive attack produced by molten salts, nevertheless the thermal treatment and procedures to be applied need to be improved in order to avoid an expensive increment in the materials cost.

Gomez-Vidal et al. [35] developed coating composed by MCrAlX (M: Ni, and/or Co; X: Y, Hf, Si, and/or Ta) to protect turbine blades from molten salt attack. Research in turbine-blade technology has reported that these coatings rely on the formation of dense and adherent alumina layers to provide high-temperature corrosion resistance against molten salts.

High-temperature corrosion-resistant coatings based on Al have also demonstrated excellent results in other sectors such as gas turbines, chemical and petrochemical plants, etc., in which their components are exposed to very harsh environments, including molten salts that are very corrosive. Audigie et al. [36] evaluated the corrosion behaviour of these slurry aluminide coatings on P92 steel at 550 and 580 °C employing industrial grade Solar Salt, obtaining that slurry aluminide coatings can significantly increase the lifetime of ferritic steels exposed to the storage temperature conditions.

The use of coating in turbine blades and some specific component of the CSP plant could be an interesting option as corrosion mitigation.

References

1. Mehos M, Turchi C, Vidal J, Wagner M, Ma Z, Ho C, Kolb W, Andraka C, Kruizenga A (2017) Concentrating solar power Gen3 demonstration roadmap. Technical Report, NREL/TP-5500-67464
2. Bradford SA (1987) Corrosion. ASM Handbook, Metals Park
3. Fernandez AG, Galleguillos H, Perez FJ (2014) Corrosion ability of a novel heat transfer fluid for energy storage in CSP plants. Oxid Met 82:331–345
4. Fernández AG, Lasanta MI, Pérez FJ (2012) Molten salt corrosion of stainless steels and low-Cr steel in CSP plants. Oxid Met 78:329–348
5. International, ASTM G1-03 (2003) Standard practice for preparing, cleaning, and evaluating corrosion test specimens. Annual Book of ASTM standards, Section 3, Metal test methods and analytical procedures, American Society for Testing and Materials (ASTM)
6. Dignam MJ, Huggins DA (1967) A kinetic study of the heterogeneous reactions of metallic sodium with chlorine and bromine. J Electrochem Soc 114(2):117–123
7. Prieto C, Gallardo J, Ruiz FJ, Barreneche C, Martinez M, Segarra M, Fernandez AI (2016) Study of corrosion by dynamic gravimetric analysis (DGA) methodology. Influence of chloride content in solar salt. Sol Energy Mater Sol Cells 157:526–532
8. Yang L (2008) Techniques for corrosion monitoring. Woodhead Publishing, Cambridge
9. Lvovich VF (2012) Impedance spectroscopy: applications to electrochemical and dielectric phenomena, 1st edn. Wiley, Hoboken, NJ
10. Cottis RA (2001) Interpretation of electrochemical noise data. Corrosion 57(3):265–284
11. ISO/FDIS (2015) Corrosion of metals and alloys—test method for high temperature corrosion testing of metallic materials by immersing in molten salt or other liquids under static conditions
12. Dorcheh AS, Durham RN, Galetz MC (2016) Corrosion behavior of stainless and low-chromium steels and in 625 in molten nitrate salts at 600 °C. Sol Energy Mater Sol Cells 144:109–116
13. Gurr M, Bau S, Pfeiffer W (2015) Investigation of the corrosion behavior of NiVAl multilayer coatings in hot salt melts. Surf Coat Technol 279:101–111
14. Novello F, Dedry O, De Noose V, Lecomte J (2014) High temperature corrosion resistance of metallic materials in harsh conditions. In: Proceedings of the 10th conference on materials for advanced power engineering
15. Walczak M, Pineda F, Fernández AG, Mata-Torres C, Escobar RA (2018) Materials corrosion for thermal energy storage systems in concentrated solar power plants. Renew Sustain Energy Rev 86:22–44
16. Lázaro A, Zalba B, Bobi M, Castellón C, Cabeza LF (2006) Experimental study on phase change materials and plastics compatibility. AIChE J 52:804–808
17. Krishna DJ, Shinde A (2017) Step by step methodology for the assessment of metal corrosion rate with PCMs suitable for low temperature heat storage applications. Mater Today Proc 4:10039–10042

18. Browne MC, Boyd E, McCormack S (2017) Investigation of the corrosive properties of phase change materials in contact with metals and plastic. Renew Energy 108:555–568
19. Ferrer G, Solé A, Barreneche C, Martorell I, Cabeza LF (2015) Corrosion of metal containers for use in PCM energy storage. Renew Energy 76:465–469
20. Sari A, Kaygusuz K (2003) Some fatty acids used for latent heat storage: thermal stability and corrosion of metals with respect to thermal cycling. Renew Energy 28:939–948
21. Cabeza LF, Illa J, Roca J, Badia F, Mehling H, Hiebler S, Ziegler F (2001) Middle term immersion corrosion tests on metal-salt hydrate pairs used for latent heat storage in the 32 to 36 °C temperature range. Mater Corros 52:748–754
22. Cabeza LF, Roca J, Nogués M, Mehling H, Hiebler S (2002) Immersion corrosion tests on metal-salt hydrate pairs used for latent heat storage in the 48 to 58 °C temperature range. Mater Corros 53:902–907
23. Cabeza LF, Roca J, Nogués M, Mehling H, Hiebler S (2005) Long term immersion corrosion tests on metal-PCM pairs used for latent heat storage in the 24 to 29 °C temperature range. Materi Corros 56(1)
24. García-Romero A, Delgado A, Urresti A, Martín K, Sala JM (2009) Corrosion behaviour of several aluminium alloys in contact with a thermal storage phase change material based on Glauber's salt. Corros Sci 51:1263–1272
25. Sastri VS, Ghali E, Elboujdaini M (2007) Corrosion, prevention and protection. Practical solutions. Wiley, New Jersey
26. Moreno P, Miró L, Solé A, Barreneche C, Solé C, Martorell I, Cabeza LF (2014) Corrosion of metal and metal alloy containers in contact with phase change materials (PCM) for potential heating and cooling applications. Appl Energy 125:238–245
27. Farrell AJ, Norton B, Kennedy DM (2006) Corrosive effects of salt hydrate phase change materials used with aluminium and copper. J Mater Process Technol 175:198–205
28. Cabeza LF, Illa J, Roca J, Badia F, Mehling H, Hiebler S, Ziegler F (2001) Immersion corrosion tests on metal-salt hydrate pairs used for latent heat storage in the 32 to 36 °C temperature range. Mater Corros 52:140–146
29. ASTM (2014) ASTM G61-14 Standard test method for conducting cyclic potentiodynamic polarization measurements for localized corrosion susceptibility of. Annual Book of ASTM Standards, pp 1–5
30. Calabrese L, Caprì A, Fabiano F, Bonaccorsi L, Borsellino C, Proverbio E (2016) Electrochemical behaviour in synthetic saliva of silane coated Ni/Cu/Ni Nd-Fe-B magnet for dentistry applications. Mater Corros 67(5):484–494
31. Natishan PM, O'Grady WE (2014) Chloride ion interactions with oxide-covered aluminum leading to pitting corrosion: a review. J Electrochem Soc 161(9):C421–C432
32. Ma H, Chen S, Niu L, Zhao S, Li S, Li D (2002) Inhibition of copper corrosion by several schiff bases in aerated halide solutions. J Appl Electrochem 32(1):65–72
33. Xu B, Li P (2014) Extending the validity of lumped capacitance method for large Biot number in thermal storage application. Sol Energy 105:71–81
34. Gibilaro M, Massot L, Chamelot P (2015) A way to limit the corrosion in the Molten Salt Reactor concept: the salt redox potential control. Electrochim Acta 160:209–213
35. Gomez-Vidal JC, Noel J, Weber J (2016) Corrosion evaluation of alloys and MCrAlX coatings in molten carbonates for thermal solar applications. Sol Energy Mater Sol Cells 157:517–525
36. Audigié P, Bizien N, Baráibar I, Rodríguez S, Pastor A, Hernández M, Agüero A (2017) Aluminide slurry coatings for protection of ferritic steel in molten nitrate corrosion for concentrated solar power technology. AIP Conf Proc 1850:0700021–0700028

Part III
Experimental Characterization of Systems for Thermal Energy Storage

Experimental Characterization of Latent Thermal Energy Storage Systems

Valeria Palomba, Andrea Frazzica, Jaume Gasia and Luisa F. Cabeza

Abstract In the present chapter, the experimental methods employed for the characterisation of latent TES are discussed. The investigated systems comprise both static and dynamic PCM storage concepts. In particular, experimental approaches applied by different laboratories are discussed in details and examples are provided to confirm the reliability of the proposed methods.

1 Introduction

Activity on latent thermal energy storage systems is vast and spread—worldwide and in time—following the diverse applications that PCM can cover. Moreover, while at materials levels some efforts are being made in order to define standard procedures for the characterization of PCMs, the different necessities, peculiarities and stakeholders of the complete system have, up to now, limited the definition of common procedures and standardization.

Within the present chapter, an overview of experimental methodologies successfully applied for the evaluation of system performance is given, focusing on the different range of application covered by latent TES, highlighting the apparatuses available and the procedures followed, in terms of methodology and calculated parameters.

Among the different systems reported in literature, focus was centred on the testing rigs employed for exemplary systems (e.g. systems with dynamic PCM) or fully exploited for the evaluation of performance and comparison of different

V. Palomba (✉) · A. Frazzica
CNR-ITAE, Istituto di Tecnologie Avanzate per l'Energia "Nicola Giordano", Via Salita S. Lucia sopra Contesse 5, 98126 Santa Lucia, Messina, Italy
e-mail: valeria.palomba@itae.cnr.it

J. Gasia · L. F. Cabeza
GREiA Research Group, INSPIRES Research Centre, Universitat de Lleida, Pere de Cabrera s/n, 25001 Lleida, Spain

A. Frazzica and L. F. Cabeza (eds.), *Recent Advancements in Materials and Systems for Thermal Energy Storage*, Green Energy and Technology,
https://doi.org/10.1007/978-3-319-96640-3_11

storages (e.g. the pilot plant at the University of Lleida or the latent storage rig at CNR-ITAE), for which details about the experimental campaign and equipment used was available.

2 Classification of Experimental Systems

A general classification of experimental systems was proposed in [1] by distinguishing "Static PCM" and "Dynamic PCM" concepts: while in the latter case, the storage material itself is recirculated in the storage, static systems do not require moving of the PCM. Static PCM systems—more common at the current development level—can be further sorted taking into account the specific function of the storage or the heat source needed to drive it. An overview of the systems that will be described within the present chapter is given in Fig. 1: "static PCM" systems are classified in multipurpose systems, making use of different heat source for general

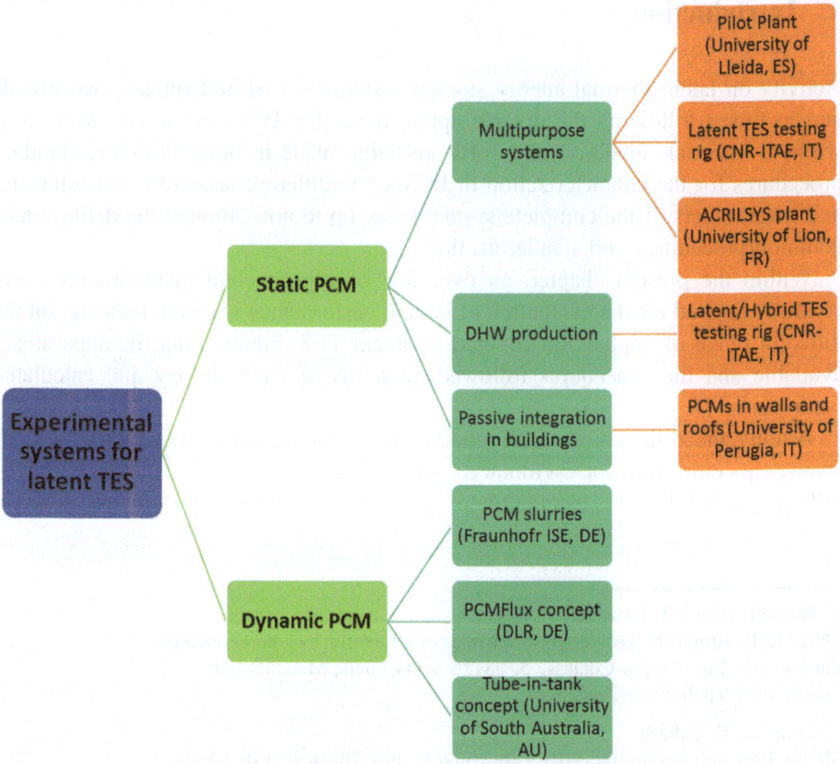

Fig. 1 Classification of the experimental systems covered in the chapter

heating or cooling purposes, systems for DHW production and systems where PCMs are part of the building, by embedding in walls or roofs. Such a classification allows distinguishing the main features required for the purpose-specific methodology needed in the evaluation of system performance and characteristics.

3 Experimental Methods with Static PCM

3.1 Pilot Plant at UdL

Activities on latent storages at University of Lleida are focused on different scales, from material development to the testing of systems in relevant size. Indeed, a characterization at pilot plant scale allows evaluating the performance of PCM at industrial scale and better extrapolating their behaviour in a real commercial plant. With this aim, a pilot plant systems was developed and realized for the testing of PCMs with medium to high melting range (up to 380 °C). One of the key features of the pilot plant is the possibility to easily test different materials and heat exchangers configurations. The characterization is done on terms of temperature profiles and power/energy rates during charging and discharging processes.

3.1.1 Description of the Experimental Setup

The experimental setup available at the University of Lleida is made up of three main systems: the heating system, the cooling system and the storage system (Fig. 2). The heating system consists of a 24 kW_e electrical boiler which is able to heat up the HTF up to 380 °C to simulate a wide range of energy sources. The cooling system contains a 20 kW_{th} air-HTF heat exchanger which cools down the HTF with the objective of simulating the energy sink. By modifying the air flow of the heat exchanger and the set-point of the heating system, different heat sinks can be simulated. Finally, the storage system to be tested includes several storage tanks of different sizes which can store the thermal energy in the sensible and latent heat form. However, in this chapter only the latent heat TES systems tested are presented. Two 0.183 m^3 storage tanks based on the shell-and-tube heat exchanger concept are used as latent heat TES system (Fig. 3). The only difference between them is the implementation of 196 squared fins in one of them, which reduces the packing factor from 0.84 (tank without fins) to 0.80 (tank with fins) and thus, the total PCM capacity. All systems are connected with stainless steel pipes which are insulated with rock wool and that are responsible for the HTF (silicone fluid Syltherm 800) distribution within a flow rate ranged between 0.3 and 3 m^3/h.

Fig. 2 Pilot plant facility available at the University of Lleida. (1) Heating system, (2) Cooling system, (3) Latent heat TES system, (4) HTF-molten salts heat exchanger, (5) Two-tank molten salts TES system

3.1.2 Operation of the Experimental Setup

Typical charge-storage-discharge cycles are performed in the experimental setup. However, due to the characteristics of the setup, a homogenization process needs to be performed before the charge and the discharge to stabilize the PCM and HTF at the desired working conditions. The characterization methodology is based on charging and discharging the PCM in a temperature range that should be at least 20 °C over and below the melting temperature of the material. Moreover, the HTF mass flow rate may be controlled. Before starting the charging process, a warming process is required to ensure that the temperature of the PCM is homogeneous at the initial temperature of charging. Once the PCM reached the initial temperature conditions for charging, the HTF is heated up outside the tank up to the inlet HTF temperature and afterwards the charging process starts. The charging process is considered to be finished once the difference between the inlet and outlet HTF temperature is <2 °C and the weighted average temperature of the middle temperature sensors in the PCM reaches the target temperature. The different steps of the tests are reported in Fig. 4.

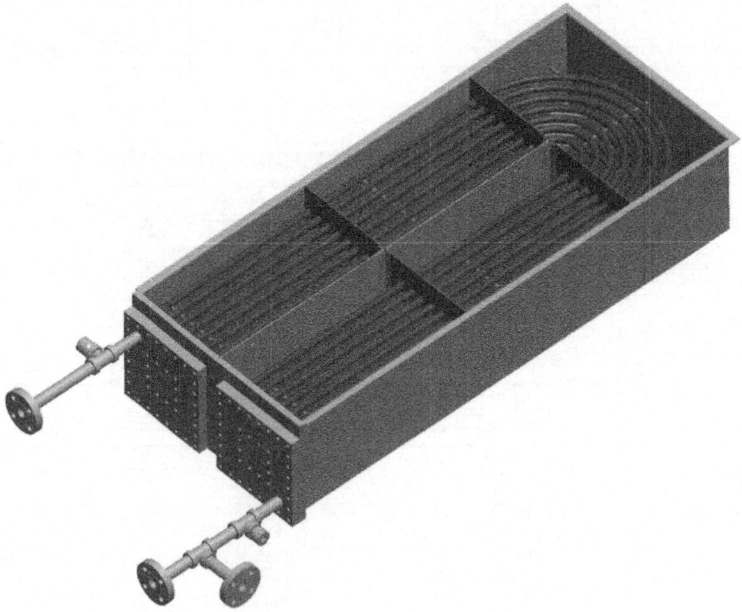

Fig. 3 Overview of the latent heat TES system used in the pilot plant facility available at the University of Lleida

3.1.3 Sensors Employed and Data Acquisition

Thirty-one PT-100 with an accuracy of ±0.1 °C are used to have a precise map of temperatures during charging and discharging processes within the PCM, together with two temperature sensors PT-100 that were installed at the inlet and outlet of the HTF tubes bundle to measure the inlet and outlet HTF temperature (Fig. 5). All the temperature sensors and HTF flow metres are connected to a data acquisition system, which controls, measures and records the information at a time interval of 30 s [2, 3].

3.1.4 Studies Performed and Results

Several studies have been performed in the pilot plant facility in order to evaluate the thermal behaviour of different PCMs under the variation of the mass flow rate, inlet temperature and the heat transfer surface. RT58 (commercial paraffin with a melting temperature of 58 °C), bischofite (a by-product from the mining industry with a melting temperature of) 99 °C, d-mannitol (a sugar alcohol with a melting temperature of 167 °C), and hydroquinone (an aromatic compound with a melting temperature of 173 °C) were evaluated and compared with the results obtained at laboratory scale.

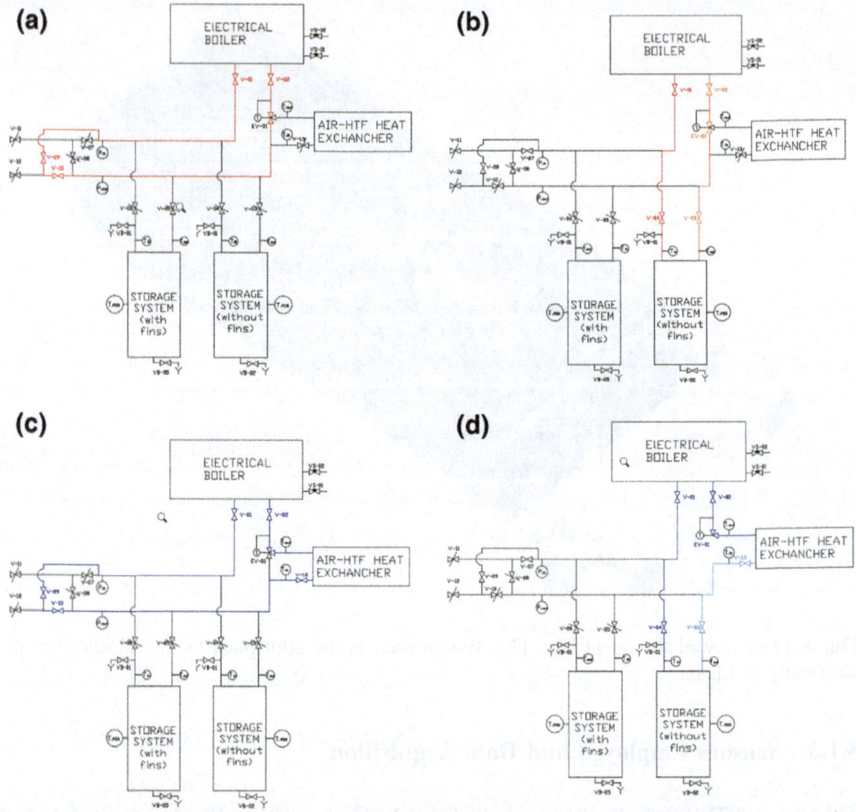

Fig. 4 Operation of the pilot plant facility available at the University of Lleida. **a** Homogenization process prior to charge. **b** Charge. **c** Homogenization process prior to discharge. **d** Discharge

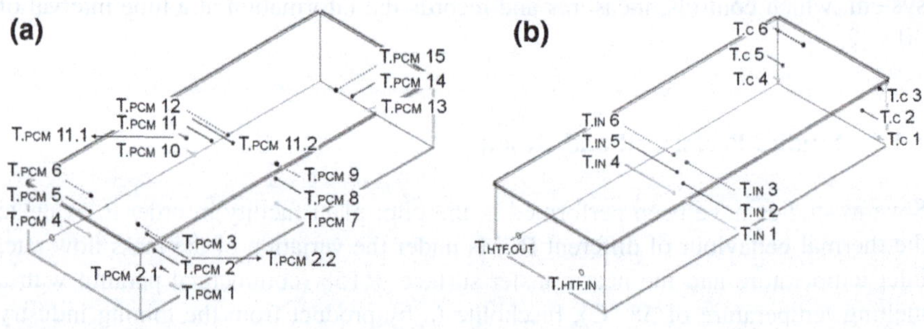

Fig. 5 Temperature sensors location. **a** Within the tubes bundle. **b** Corners and central part of the tanks

An example of the results of the experiments carried out in such installation with RT58 is presented in Fig. 6. With this figure, the charging and discharging processes can be evaluated and compared. In the example shown it was observed that the discharge was 42% slower than the charge. The main reasons for these differences in time lie in the presence of natural convection during the melting process, which enhanced the charging process and the creation of a solid layer around the tubes bundle when the PCM started to solidify, which hindered the discharging process. Other parameters that can be evaluated are the power profiles and the energy profiles.

3.2 *Static PCM Concept at CNR-ITAE: Multipurpose Testing Rig for Latent Heat Storages*

Activities on latent storages at CNR-ITAE aimed at covering the various levels required to bring the concept of an innovative storage (TRL 3) into a computer-aided prototype design (by means of numerical analysis and CAD design) and its laboratory testing under controlled conditions (TRL 5). A testing rig specifically suited for latent TES in the range of 50–95 °C was developed, simulating solar heat or waste heat sources and space heating distribution systems or thermally activated chillers. It allows easily replacement of the heat exchanger and is suitable for testing storages up to discharge power of 20 kW, which is a scale close to real applications.

3.2.1 Description of the Experimental Setup

In the design of the testing rig, the following challenges were specifically addresses:

- flexibility: the testing rig should allow the testing of different systems under the widest range of operating conditions possible;

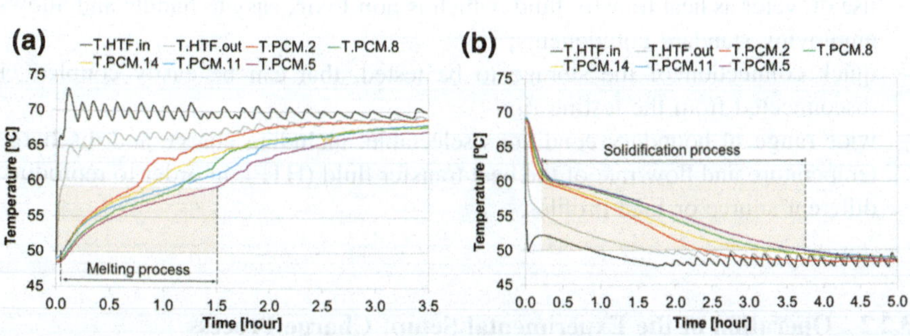

Fig. 6 PCM and heat transfer fluid (HTF) temperature profile obtained with the pilot plant available at the University of Lleida: **a** charging process and **b** discharging process [2]

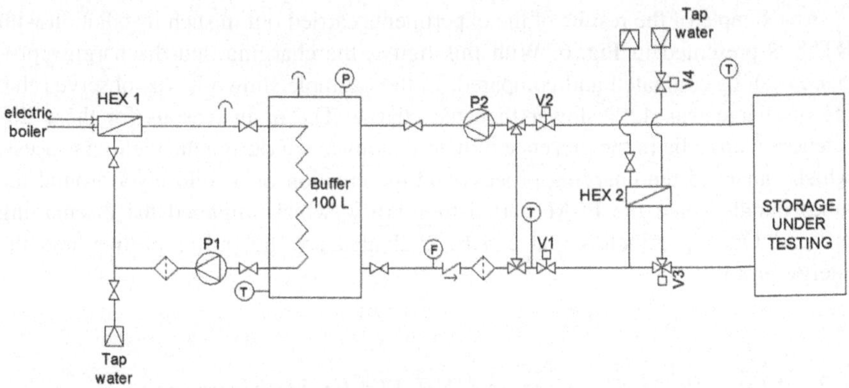

Fig. 7 P&ID of the testing rig at CNR-ITAE [4]

- automatic operation: the rig should allow for remote control, without the need for an operator to control the charge/discharge;
- real-time acquisition of all the relevant parameters: all the transducers employed should be connected to a suitable data-logging system, to record and control all the main parameters in real time.

The P&ID of the testing rig is shown in Fig. 7. The main core of the rig is the tested storage that is connected to a second service storage (called "Buffer" in Fig. 7). The buffer is equipped with an internal heat exchanger and is connected to a 24 kW diathermic oil heater through a plate heat exchanger ("HEX1" in Fig. 7). In order to allow the operator to set an arbitrary discharge temperature, a second plate heat exchanger ("HEX2" in Fig. 7) is connected to the return pipe from the buffer and to tap water. A motorized mixing valve is used to control the temperature of the water exiting from HEX2. Such a valve works according to a PID algorithm implemented in LabVIEW. Each component of the testing bench is isolated by using two-way valves, in order to facilitate maintenance and replacement. A picture of the testing rig is shown in Fig. 8. The main characteristics of the testing rig are:

- use of water as heat transfer fluid, which is non-toxic, easy to handle and allows employing standard components;
- quick connection of the storage to be tested, that can be easily coupled or disconnected from the testing rig;
- wide range of boundary conditions selectable, including charge and discharge temperature and flow rate of the heat transfer fluid (HTF), in order to reproduce different source or load profiles.

3.2.2 Operation of the Experimental Setup: Charge Process

The charging of the storage to be tested is realized in three steps:

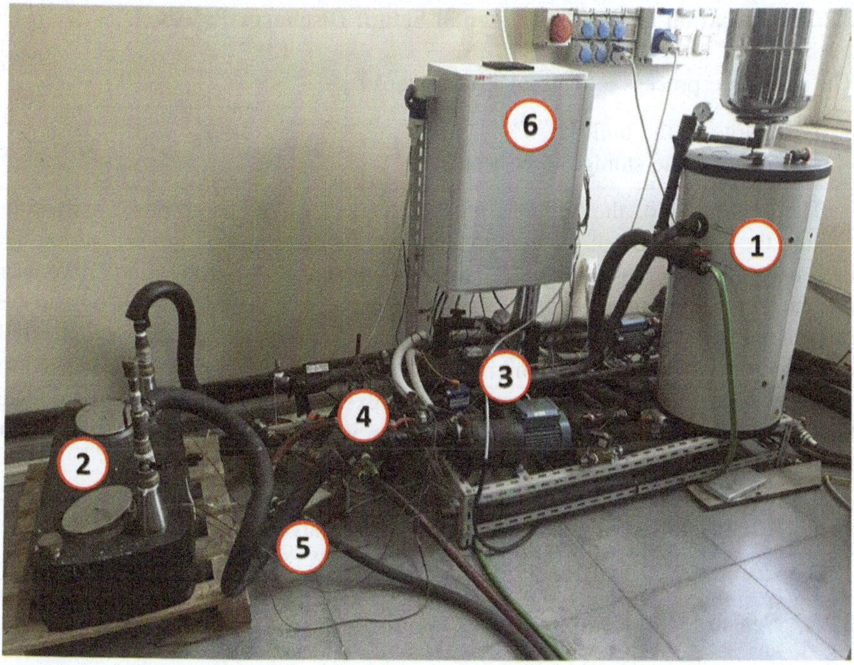

Fig. 8 Picture of the testing rig at ITAE. (1) service buffer; (2) PCM storage under testing; (3) hydraulics; (4) mixing valve; (5) plate heat exchanger for discharge; (6) electronics

- filling of the whole hydraulic pipes and the storages, thanks to the connection to tap water;
- pre-charge of the buffer storage, that is heated up to the user-selected temperature thanks to the connection with the heater;
- charge of the latent or hybrid storage, by opening the connection with the buffer storage.

During the last two steps, the hydraulic pump P1 (as called in Fig. 7) is used to circulate the HTF in the buffer. The final temperature of the buffer can be set on the diathermic oil heater in the range 50–99 °C. The pump has a head of 5.5–20.5 m and a maximum flow rate of 4.8 m^3/h. The plate heat exchanger HEX1 is a brazed plate heat exchanger, with plates in AISI316L, chosen in order to tolerate the temperature and pressure conditions required during operation. A by-pass pipe, in which the HTF is circulate with the pump P2, was installed, with the aim of speeding up the pre-charge of the buffer.

During the charge of the storage, the pump P2 is used for the circulation of the fluid between the buffer and the storage under testing. The connection is open or closed thanks to the solenoid valves V1 and V2 that are controlled remotely. The pump P2 has a maximum flow rate of 4.8 m^3/h and a maximum head of 42.5 m. Moreover, on both the storages, safety valves were installed to avoid dangerous pressures.

3.2.3 Operation of the Experimental Setup: Discharge Process

The discharge process is realized in two steps:

- pre-charge of the buffer;
- discharge of the storage to be tested.

The pre-charge of the buffer is realized as during the charge process, by controlling the connection of the component with the diatermic oil heater. During the discharge, instead, the desired discharge temperature is set on the LabVIEW user interface. At each acquisition step, the set is compared to the inlet temperature to the storage under testing. According to the difference between the two values, a 0–10 V command signal is sent to the mixing valve V3. The mixing valve is connected to the inlet pipe of the storage to be tested and mixes the water coming from the plate heat exchanger HEX2 that is connected to tap water, and the circuit of the buffer storage, that is at high temperature.

The chosen mixing valve is a ball valve, with modulating motor and analogue signal. The maximum differential pressure of the valve is 350 kPa.

3.2.4 Sensors Employed and Data Acquisition

The P&ID of Fig. 7 shows also the sensors chosen and installed in the testing rig:

- class A T-type thermocouples, used for the measurement of the temperature of the water inside the buffer and of the PCM inside the storage to be tested.
- Pt100 1/10 DIN with four wires, used for the measurement of inlet and outlet temperature of the HTF. The high accuracy of these sensors allows a precise evaluation of the energy accumulated by the system. Before the installation, the sensors were calibrated.
- MVM040 magnetic flow metre with 0.25% FS accuracy, used to measure the flow rate of the HTF.

Furthermore, the flow rate of the pump P2 is controlled through a variable-speed drive, in order to guarantee flexibility and allow analysing the effect of such a parameter on the storage capacity or power.

All the sensors and electric components (pumps and valves) are connected to a National Instruments FieldPoint acquisition system, which is composed of a controller and I/O modules. A specifically developed LabVIEW VI is used to assure the management of the tests (i.e. the opening/closing of the valves, the on/off of the pumps and the control of the mixing valve), the acquisition and the monitoring of all the relevant quantities.

3.2.5 The Tested Systems

The developed testing rig was used for the characterization of two storage systems, employing the same PCM, a commercial paraffin by PCMProducts with the commercial name of PlusICE A82, with nominal melting temperature of 82 °C. Since the main challenge in the exploitation of paraffins for heat storage purposes is their low thermal conductivity, the chosen configurations are aimed at increasing heat transfer. In particular, enhancement at macroscale, by augmenting exchanger area was the strategy pursued. Consequently, the realized storage systems consist of:

- A customized fin-and-tube HEX, having overall volume of 50 l, with fin distance of 5 mm, and tubes organized in four ranks, with eight pipes in series per rank
- A commercial asymmetric plate HEX, having overall volume of 32 l, with 16 l on the PCM side and distance between the plates of 3.5 mm.

The main features of the prototypes are reported in Table 1, while in Fig. 9 the two systems are depicted: on the left, the fin-and-tube HEX based storage and on the right, the plate HEX based one.

Figure 10, instead, shows the position of temperature sensors in the two systems. It is worth noticing that the leading criterion in locating the temperature sensors was the possibility to identify longitudinal and transversal gradients.

Table 1 Main features of the tested storages

	Fin-and-tube heat exchanger	Plate heat exchanger
PCM type	Plus ICE A82	Plus ICE A82
PCM mass	37 kg	16 kg
Construction material	ASTM316L, copper brazing	AISI 416
Overall dimensions	400 × 650 × 350 mm	727 × 320 × 166 mm
Overall mass (PCM included)	239 kg	57 kg
Number of pipes	32	
Diameter pipes	½″ internal pipes, 1 ½″ inlet/outlet connections	1 ½″ inlet/outlet connections
Ranks	4	–
Number fins	48	–
Fin space	5 mm	–
Number of plates	–	40
Plate space	–	3.5 mm
Heat transfer fluid	Water	Water

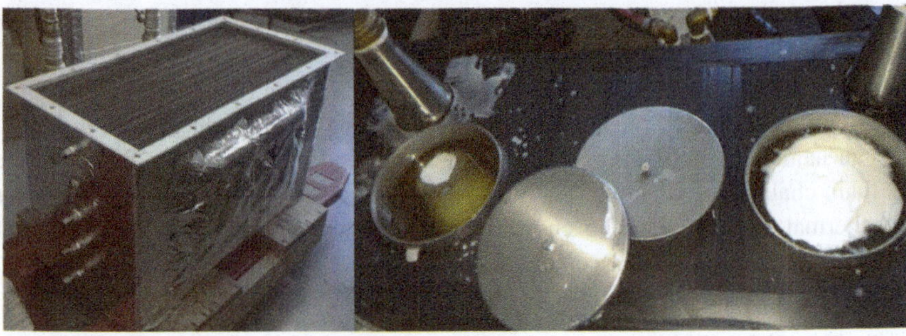

Fig. 9 The heat exchangers tested at ITAE

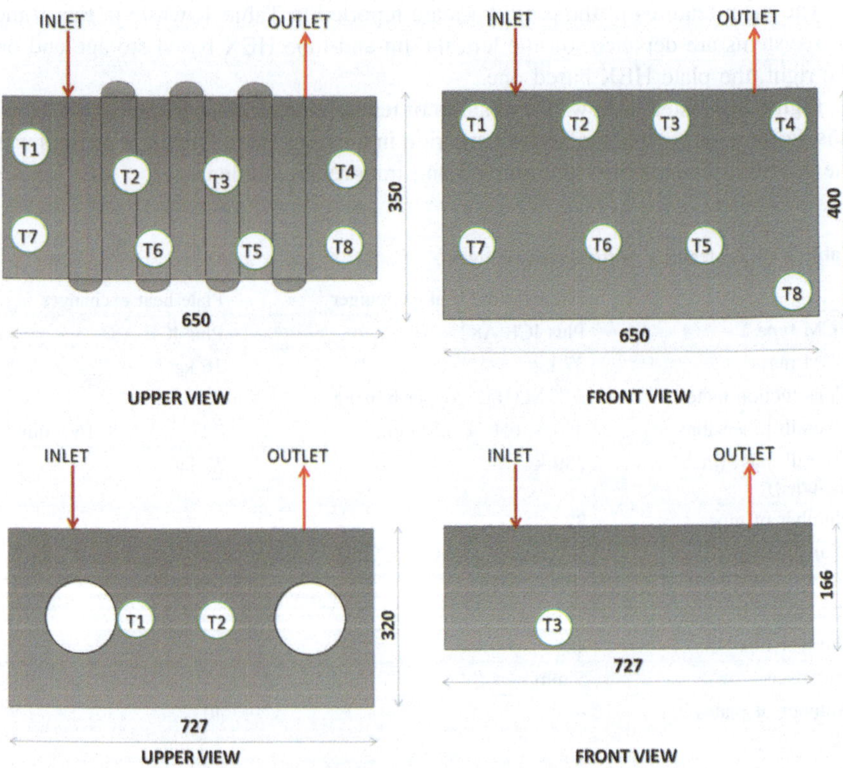

Fig. 10 Position of temperature sensors in the storages investigated

3.2.6 Experimental Procedures

Charge procedure is realized as follows:

1. Definition of target charge inlet and final temperatures
2. Connection of the heater to the service buffer until the temperature of the buffer is equal to $T_{0,\mathrm{charge}} + 5$ °C
3. Opening of the connection between the service buffer and the storage to be tested
4. The charge is stopped when the average temperature in the storage to be tested is equal to the target one.

Discharge procedure is realized as follows:

1. Definition of target discharge inlet temperature
2. Connection of the heater to the service and of tap water to the plate heat exchanger
3. Setting, on user interface of the control programme, of desired discharge temperature
4. Opening of the connection between the storage to be tested and the discharge loop

Discharge is stopped when average power is lower than 100 W.

The following figures were calculated for characterization and comparison purposes:

- Heat transfer rate between the HTF and the PCM:

$$\dot{Q}_{\mathrm{HTF}} = \dot{m}_{\mathrm{HTF}} c_{\mathrm{p}} \left(T_{\mathrm{HFT,in}} - T_{\mathrm{HFT,out}}\right) \tag{1}$$

where, $\dot{Q}_{\mathrm{HTF}}$ is the heat transfer rate, $\dot{m}_{\mathrm{HTF}}$ is the HTF mass flow rate and $T_{\mathrm{HFT,in}}$, $T_{\mathrm{HFT,out}}$ are the HTF inlet and outlet temperatures.

- Energy charged/discharged, calculated as the integral of the instantaneous power for the whole duration of the charge/discharge process.

$$\Delta E_{\mathrm{PCM}} = \int_{0}^{\tau_{\mathrm{fin}}} \dot{m}_{\mathrm{HTF}} c_{\mathrm{p}} \left(T_{\mathrm{HFT,in}} - T_{\mathrm{HFT,out}}\right) \mathrm{d}\tau \tag{2}$$

- Charge efficiency, calculated as the theoretical energy to be supplied for the charging process over the measured one. The theoretical energy was obtained by adding the contribution of the PCM (from the integral heat curve) and the sensible heat of the metal masses of the storage.

$$\varepsilon_{ch} = \frac{\Delta E_{PCM,th}}{\Delta E_{PCM,ms}} \tag{3}$$

where $\Delta E_{PCM,th}$ is the theoretical energy and $\Delta E_{PCM,ms}$ is the measured one during the process.

- Discharge efficiency, calculated as the total amount of energy recovered from the system over the theoretical one.

$$\varepsilon_{disch} = \frac{\Delta E_{PCM,ms}}{\Delta E_{PCM,th}} \tag{4}$$

3.2.7 Results

Figures 11 and 12 show the temperatures measured in different parts of the heat exchangers during the tests, as well as the temperatures of the HTF: Fig. 11 refers to the storage employing the custom fin-and-tube heat exchanger, Fig. 12 to the storage employing the plate heat exchanger. Indeed, what is clear is that the temperatures in the plate heat exchanger are more uniform whereas in the fin-and-tube heat exchanger significant gradient, especially in the horizontal direction is highlighted. In both cases, the leading heat transfer mechanism identified is conduction. However, due to the poor thermal conductivity of the PCM employed, the material that is far from heat transfer surfaces (the fins in one system, the plates in the other) undergoes the phase change really slowly. As it is possible to notice from the pictures, both charge and discharge are faster in the plate heat exchanger.

Another important observation is that the chosen material presents a wide melting/solidification interval and therefore it is not possible identifying, from the temperature profiles reported, the phase change interval in a clear way. Instead, after an initial rapid temperature increase, corresponding to the sensible heating of the

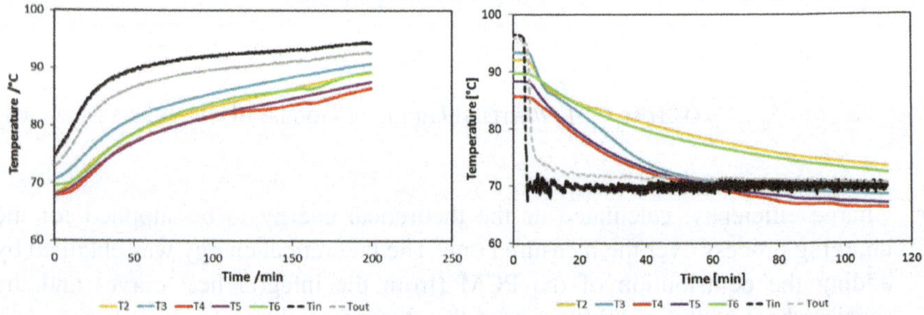

Fig. 11 Temperatures in the storage with the fin-and-tube heat exchanger during charge and discharge

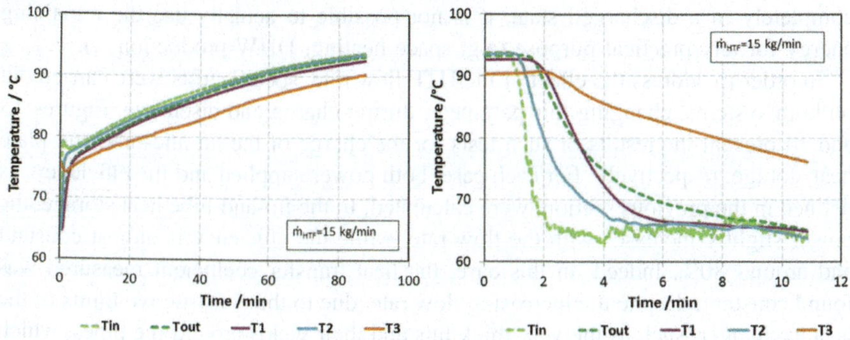

Fig. 12 Temperatures in the storage with the plate heat exchanger during charge and discharge

material, there is a constant increment in the temperature of the PCM, with a different slope than for the sensible heating of the material. This is especially visible in the profiles for discharge tests, especially for the case of the plat heat exchanger, where the phase change occurs very rapidly (<15 min).

As stated above, the actual amount of material reacting (i.e. completing the phase change) is not the entire volume of PCM inside the heat exchanger. This is especially clear in the temperature profiles for discharge: T2 and T3 in Fig. 11 and T3 in Fig. 12 remain above solidification temperature. At the same time, however, it is not possible to collect any useful effect in the HTF circuit (and therefore for the user), thus indicating that heat transfer between the material and the heat exchanger represents the main critical issue in such systems.

Effect of HTF Flow Rate

The flow rate of HTF fluid represents the main parameter on which is possible to operate in order to adjust the performance of the TES and adapt it to user requirements, in terms of power to be delivered and temperatures to be maintained. At the same time, as explained in the previous section, heat transfer between the pipes or plates of the heat exchanger and the PCM is critical. It is well known that turbulent flow of the HTF increases the heat transfer capability and efficiency of the heat exchanger. When the flow is in laminar regime, which is the case of the fin-and-tube heat storage, the importance of HTF flow rate is even higher: the higher the flow rate, the higher the heat transfer coefficient.

Furthermore, the flow rate has an influence on the temperature of the HTF, which, in turns, affects the temperature difference between the HTF and the PCM. This leads to two effects: it alters the driving force for the phase change process, represented by such a ΔT. On the other hand, it changes the useful energy collected by the user. Indeed, when the temperature difference between the inlet and outlet of the heat transfer carrier becomes too low, even though if the material is not completely in a discharged state, it is not possible to actually use the remaining energy for any practical purpose (e.g. space heating, DHW production…).

In order to address the effect of the HTF flow rate, specific tests were carried out on both systems, changing this parameter during charge and discharge. Figures 13 and 14 present the results of such tests for the charge of the fin-and-tube and plate heat storage, respectively. For each case, both power supplied and the efficiency, as defined in the previous section, were calculated. In the fin-and-tube heat storage, the power slightly increases with the flow rate, while the efficiency is almost constant and around 80%. Indeed, in this case, the heat transfer coefficient measured was found constant, despite the increasing flow rate, due to the constructive limits of the heat exchanger, such as the very thick fins and their welds around the pipes, which increase the thermal resistance. On the contrary, in the plate heat storage, increasing the flow rate improves the heat transfer inside the channels; even though the effect on the average power is still limited, the benefit is evident when calculating the efficiency of the system. Such a parameter is higher when the heat losses are lower

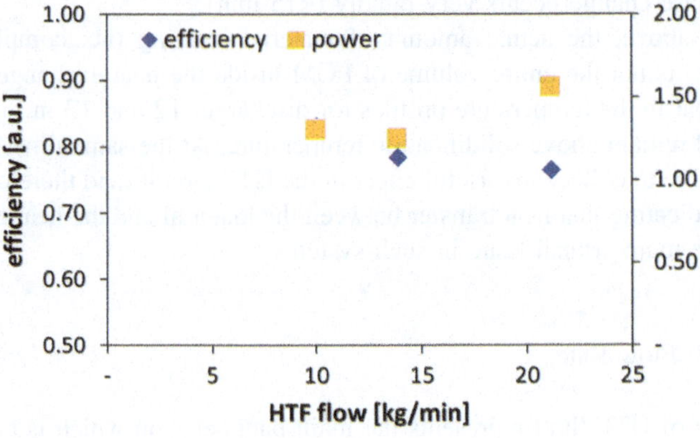

Fig. 13 Effect of HTF during charging of the fin-and-tube heat storage

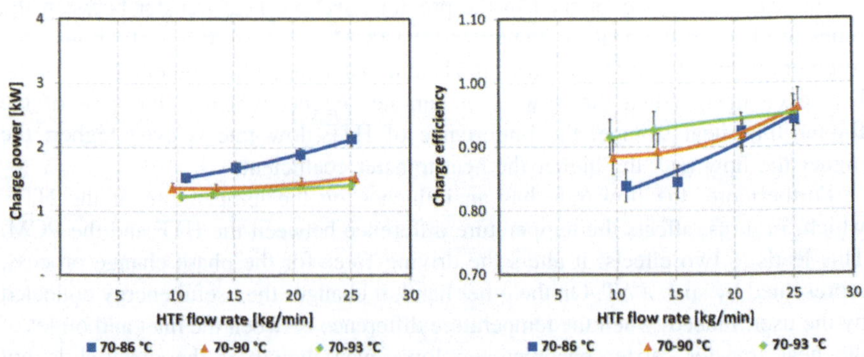

Fig. 14 Effect of HTF during charging of the plate heat storage

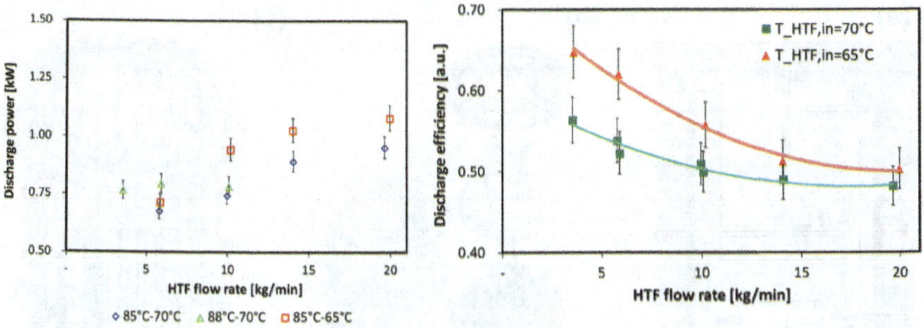

Fig. 15 Effect of HTF during discharging of the fin-and-tube heat storage

and the heat transfer between the exchanger and the PCM is enhanced. Increasing the flow rate improves the heat transfer efficiency of the heat exchanger, allowing also for a shorter charge period and improving the achievable charge efficiency.

Figures 15 and 16 show the effect of the flow rate during the discharging process, which is, from a practical point of view, more important than the charge. The tendency is opposite with respect to the previous case: increasing the flow rates increases the average power delivered to the user, but reduces the efficiency of the process. Indeed, increasing the flow rates reduces the outlet temperature from the storage, which limits the useful energy that is possible to extract from the storage. According to what reported in the literature [5, 6], the efficiency decreases according a second-order power with the increasing flow rate. There is a marked difference between the two systems tested: the average discharge power of the fin-and-tube heat exchanger is always around 1.2 kW, for the poor design and

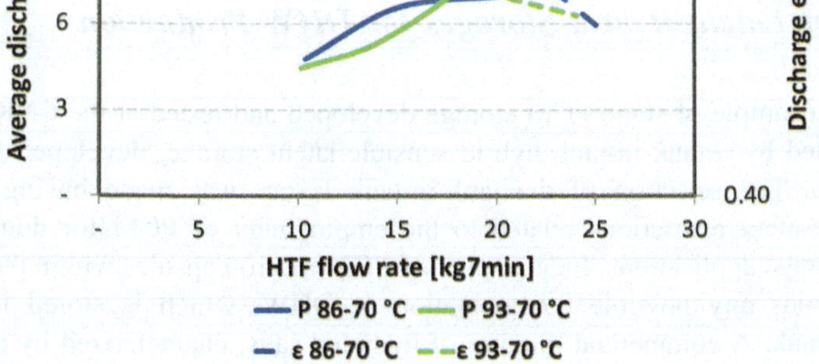

Fig. 16 Effect of HTF during discharging of the plate heat storage

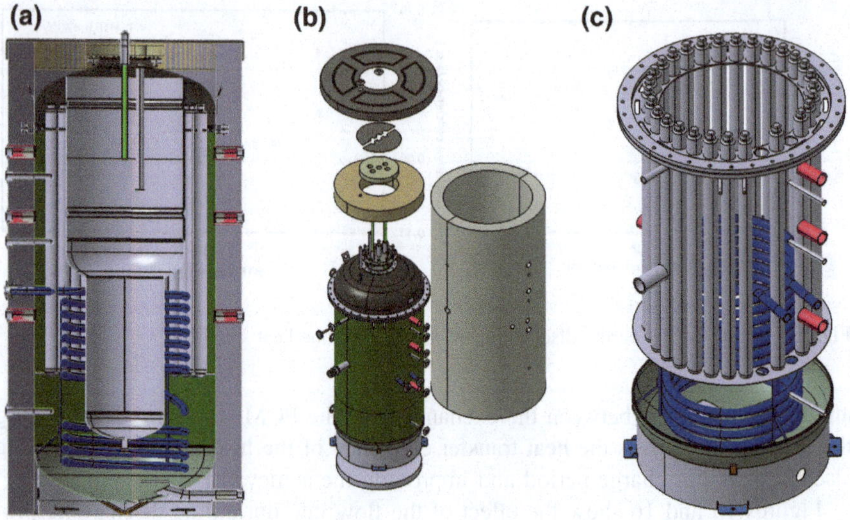

Fig. 17 Rendering of the designed tank-in-tank storage: longitudinal section (**a**), three-dimensional complete view (**b**) and three-dimensional detail of the integrated PCM macro-capsules (**c**)

construction features chosen. The efficiency lies in the range 50–70%, according to the power chosen. The discharge period is quite long, around 1 h, making the storage feasible for applications where getting a continuous effect is the main constraint. The average discharge power of the plate heat storage, instead, goes from 4 kW to 12 kW, according to the chosen flow rate. The efficiency is slightly higher, up to 80%, thus making the system competitive even for applications where high power is necessary (e.g. DHW production). A more detailed analysis on experimental results, together with the comparison between the two concepts proposed is given in [4, 7].

3.3 Static PCM Concept at CNR-ITAE: Testing Rig for Latent/Hybrid Storages for DHW Production

Another example of static PCM storage developed and tested at the CNR-ITAE is represented by a tank-in-tank hybrid sensible-latent storage, developed for DHW provision. The selection of the tank-in-tank layout was made having in mind possible safety restrictions related to the employment of PCM for domestic hot water storage applications. Indeed, placing PCM macro-capsules within the external tank, avoids any possible contamination of DHW, which is stored inside the internal tank. A commercial mixture of hydrated salts, characterized by a nominal melting temperature of 58 °C, was selected to be included inside the heat storage.

It was provided already macro-encapsulated in tubes made of HDPE. In Fig. 17, the three-dimensional drawings of the developed storage are reported. It is a standard tank-in-tank storage modified in order to include a certain amount of PCM macro-capsules in the external tank (mantle). In particular, it is able to contain up to 27 tubes, which means about 67 kg of PCM, whose overall volume is 53 l. The TES can be charged both by means of an immersed copper coil, especially for solar applications, and by direct exchange of water in the mantle, when a boiler or a CHP is employed. As the expected application of the realized TES was the installation in conjunction with a micro-CHP for domestic applications, the latter way of charging phase was simulated by means of the test rig. The installed TES in the ITAE's lab is reported in Fig. 18.

In order to experimentally characterize the TES performance, two different set of tests were performed, as already suggested in literature [8, 9] and summarized in Fig. 19:

– Test A: Multiple 5 min discharges: the storage is heated up; afterwards the DHW is withdrawn for 5 min with a subsequent stand-by period of 30 min. This procedure is repeated until the storage is able to deliver DHW at 45 °C.

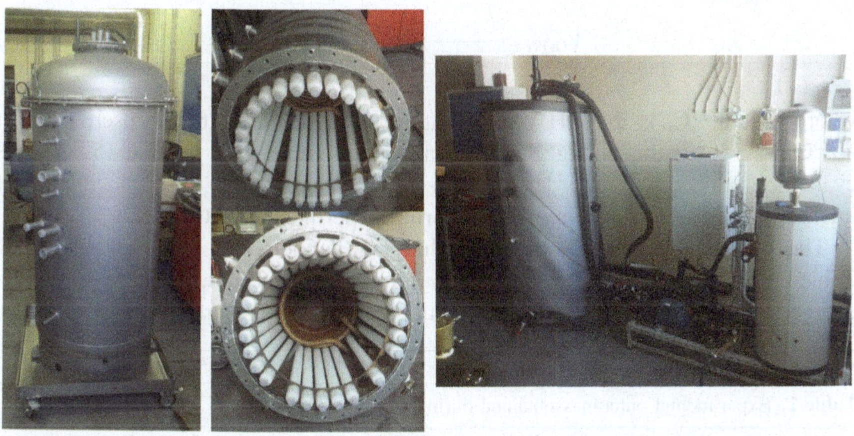

Fig. 18 Some pictures of the installed storage in the CNR-ITAE lab

Fig. 19 Details of the employed testing conditions

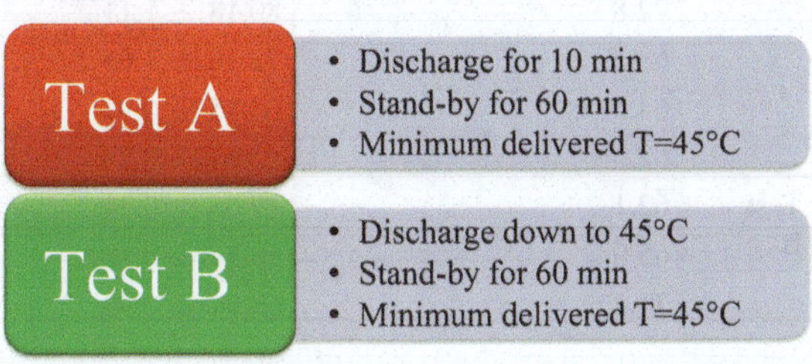

- Test B: Discharge up to 45 °C: the storage is heated up; afterwards the DHW is continuously withdrawn as long as the DHW temperature is higher than 45 °C. Subsequently the storage is put in stand-by for 30 min, and then the procedure is repeated.

The experimental activity was performed by varying the charging temperature, aiming at the evaluation of the effect of this parameter on the achievable TES density. In order to assure the completed melting of the PCM, the charging process was protracted until reaching a really low temperature difference between inlet and outlet temperature on the heating side (i.e., <0.2 K).

The discharging process was carried out by withdrawing DHW from the top of the internal tank, which was employed to increase the temperature of tap water (average flow rate 8 l/min) by means of the intermediate HEX2 highlighted in Fig. 7. The automatic mixing valve allowed setting a temperature to be delivered to the user of about 40 °C. Each test was stopped as soon as the temperature of the DHW inside the tank fell below 45 °C. In order to get a comparable performance figure, the energy delivered to the user was calculated for all the experimental tests, according to the Eq. 2.

Moreover, starting from the energy delivered, an equivalent amount of DHW produced has been calculated, according to the following Eq. 5:

$$V_{\mathrm{DHW}} = \frac{\Delta E_{\mathrm{DHW}}}{\rho_{\mathrm{w}} c_{\mathrm{p_w}} (T_{\mathrm{DHW}} - T_{\mathrm{net}})}, \tag{5}$$

where, T_{DHW}, °C, is the DHW temperature to be delivered to the user, fixed at 40 °C, T_{net}, °C, is the average seasonal tap water temperature, fixed at 15 °C: This evaluation is necessary in order to avoid the influence related to the fluctuation of inlet tap water temperature registered during experimental testing.

The obtained results for different charging temperatures are summarized in Table 2 for both Test A and Test B conditions. As expected, an increase in discharged energy is registered when the initial temperature is increased. Nevertheless,

Table 2 Experimental outcomes obtained during the Test A and Test B

	Initial tank temperature (°C)	Discharged energy (MJ)	Equivalent DHW production (dm^3)
Test A	74.5	83.92 ± 1.18	801.9 ± 11.2
	72.8	80.18 ± 1.11	766.2 ± 10.6
	68.3	72.76 ± 1.05	695.3 ± 10.0
	64.5	63.45 ± 0.91	606.3 ± 8.7
	62.5	55.57 ± 0.75	531.0 ± 7.1
Test B	62.1	47.65 ± 0.71	555.5 ± 6.8
	65.9	58.14 ± 0.80	555.5 ± 7.7
	71.4	68.97 ± 0.92	659.1 ± 8.8

such an increase is more evident for temperatures closer to the nominal PCM melting temperature, as in this way the latent heat is better exploited, thanks to the reduced temperature spread between stored DHW and cold tap water. Furthermore, it can be pointed out that the Test B conditions cause a reduction of DHW production, compared to the Test A. This can be related to a lower efficiency in exploiting the latent heat of encapsulated PCM, probably due to a low heat transfer efficiency between macro-capsules and water contained into the mantle. The experimental error was evaluated being always lower than 2% of the measured value.

Comparing the achieved TES densities for hybrid configuration, reported in Table 2, against the ones obtained for the sensible configuration, an increase ranging between 13% (at 65 °C of charging temperature) and 8.5% (at 72 °C of charging temperature) was highlighted. As expected, the closer is the charging temperature to the PCM melting temperature the higher is the TES density increase. Interestingly, the same result was obtained regardless of the testing condition (i.e. Test A or Test B), confirming the reliability and the effectiveness of the proposed experimental testing method.

3.4 Static PCM Concept at University of Lyon: ACLIRSYS Plant Testing Rig

ACLIRSYS plant located at the University of Lyon was developed with the aim of studying large scale latent heat storage systems coupled with heat pump (e.g. for HVAC systems in building application) under winter and summer conditions.

The experimental setup, described in [10], is shown in Fig. 20. Its core is a PCM storage tank, coupled to a heat pump. In particular, for the experimental campaign reported in [10], encapsulated ice/water nodules were used as PCM. Operation of the system is based on the connection of three sub-systems: the first one is a 40 kW air-water heat pump equipped with a variable-speed scroll compressor, a receiver tank, a four-way valve which allows the heat pump to alternate between a heater and a refrigerator, a refrigerant-air heat exchanger, a refrigerant-water heat exchanger and an electronic expansion valve. The refrigerant is R-410A. The heat pump is placed in an insulated test room where the temperature is regulated. The second sub-system is the PCM storage system including a variable-speed pump and a distribution device made of different electro-valves that allow the system to run in several modes, mainly: (i) direct energy production by the heat pump; (ii) storage charge; (iii) storage discharge. HTF is a water/glycol solution. Finally, three heating electrical resistances in parallel (40 kW; 20 kW; 20 kW) and three pumps allow simulating the cooling loads of buildings.

The data acquisition system consists of temperatures sensors (thermocouples) for the measurement of temperatures in the circuit of the HTF, two flow metres, refrigerant pressure transmitter and power metres to measure active and apparent

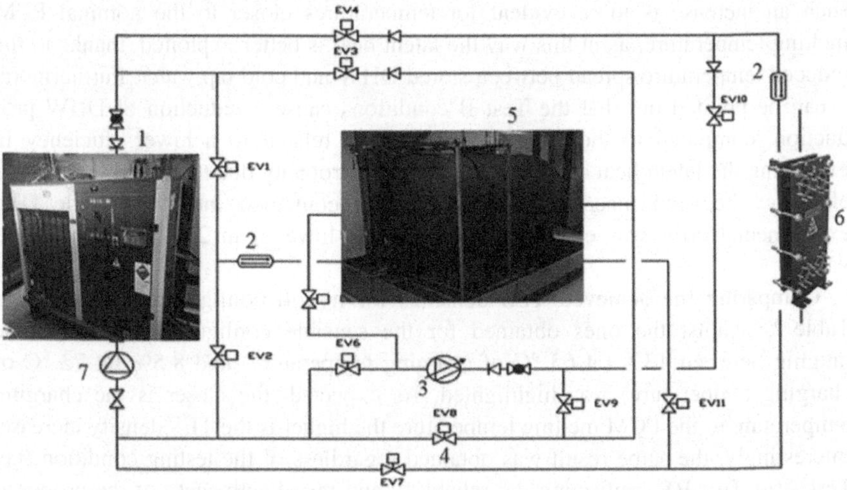

Fig. 20 Schematic diagram of the ACLIRSYS plant: (1) heat pump, (2) flow metre, (3) variable-speed pump, (4) expansion valve, (5) PCM storage tank, (6) 80 kW heat exchanger, (7) constant-speed pump [10]

powers of the compressor, overall heat pump, and variable-speed pump electrical consumption. All the input variables can be controlled by the user through the LabVIEW® interface. Only the heat pump electronic expansion valve position is automatically set by an inner controller in order to regulate the evaporator superheat.

Results of the experimental campaign reported were used to calibrate and validate a model of the system.

4 Experimental Methods with Dynamic PCM

Under the name of "Dynamic PCM" there is a variety of experimental concepts relying on the recirculation of the storage material, with the aim of increasing heat transfer inside the PCM through continuous mixing, thus reducing also segregation and limiting the degradation of the heat storage material. The idea of moving the PCM comes from applications for freezing, i.e. under 0 °C, where the useful effect to the consumer is obtained during melting of the material. Different concepts for active storage systems have been developed during the last year, applying such a method, the most relevant of which are reported in the next sections.

4.1 Dynamic Melting Concept at the University of South Australia: Tube-in-Tank System

The idea of keeping the melted PCM in movement was applied at the University of South Australia to different heat exchanger concepts. The underlying idea is to create melted paths in the frozen PCM using pre-melt tubes and, once the state of charge is advanced enough, circulating the PCM through a pump. The application of such a concept to a tube-in-tank system is described in [11]. Figure 21 shows the experimental facility used. It consists of the tube-in-tank system under testing, a 540 l tank for the HTF and a water heater. The HTF is cooled down by a refrigeration unit. In addition, pumps and ball valves allow the correct management of the fluids. The sensors installed include three flow metres, RTDs in the fluid circuits and thermocouples inside the tube-in-tank storage. The HTF fluid used is a non-combustible, aqueous based fluid with dissolved ionic solids, which can operate even below −50 °C.

The tested system is shown in Fig. 22. Results on the system proved that the average effectiveness of the system, defined according to [6] can be increased up to 89%. The effective thermal conductivity of the PCM was doubled with respect to the static case, which corresponds to the use of anular fins along the tubes. Finally, the time taken for the melting process was shortened.

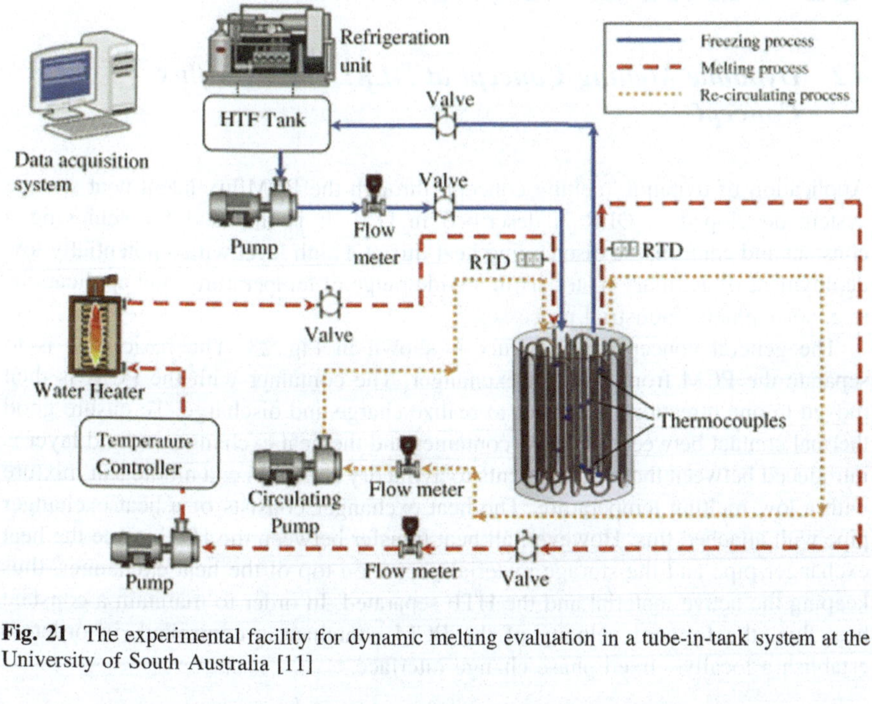

Fig. 21 The experimental facility for dynamic melting evaluation in a tube-in-tank system at the University of South Australia [11]

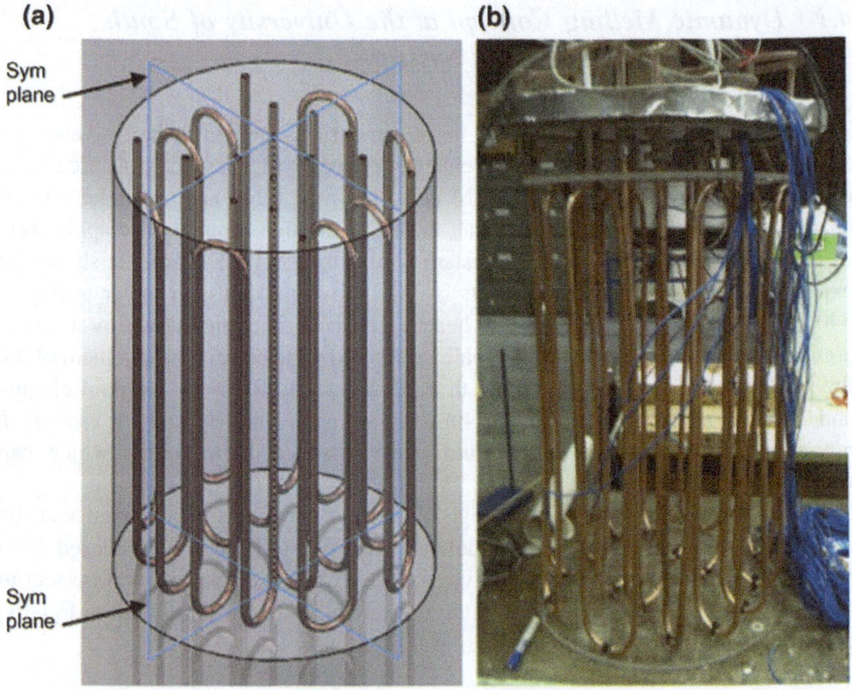

Fig. 22 The tube-in-tank system tested at the University of South Australia

4.2 Dynamic Melting Concept at DLR: The PCMflux Concept

Application of dynamic melting concept, through the PCMflux latent heat storage system developed at DLR is described in [12]. It is intended for achieving a constant and controllable discharging heat flux at a high level with a potentially low deployment of auxiliary materials in a wide range of temperatures and applications (e.g. solar plants, industrial processes).

The general concept of PCMflux is shown in Fig. 23. The basic idea is to separate the PCM from the heat exchanger. The container with the PCM is then moved in one direction or another to realize charge and discharge. To ensure good thermal contact between the PCM container and the heat exchanger, a fluid layer is introduced between these components to avoid dry contact, i.e. a nitrate salt mixture with a low melting temperature. The heat exchanger consists of a heat exchanger pipe with attached fins. However, all heat transfer between the HTF inside the heat exchanger pipe and the storage material occurs on top of the heat exchanger, thus keeping the active material and the HTF separated. In order to maintain a constant heat flux, the forward velocity of the PCM container is controlled, in order to establish a locally—fixed phase change interface.

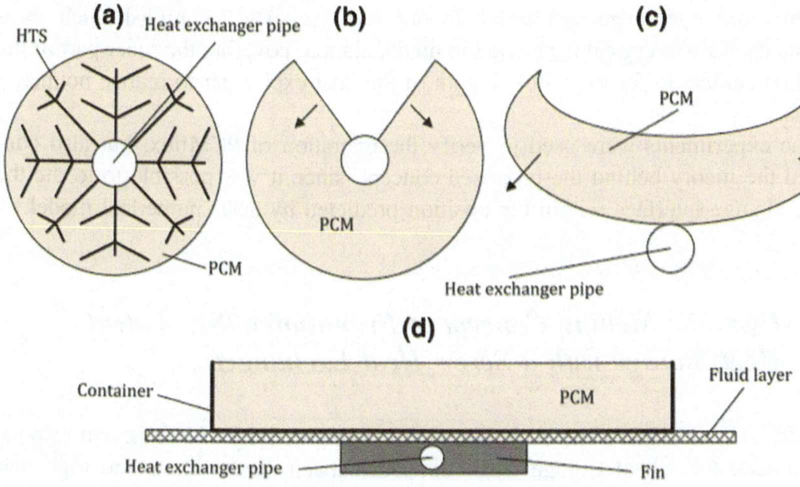

Fig. 23 The PCMflux concept starting from the state of the art with heat transfer structure (HTS) (**a**) Over the unrolling of the PCM from the heat exchanger (**b**, **c**) to a general PCMflux module (**d**) [12]

Figure 24 shows the apparatus used for the experimental validation of the proposed concept. Compared to the basic concept, the PCM is not filled into one single container, but rather into different cavities connected by wires. The wires are used to pull the containers by a transport mechanism into the requested direction. The temperatures levels for charge and discharge are ensured by two thermostatic baths employing diathermic oil as HTF and thermocouples for temperature control. Movement is realized through a drive system including a reducing gear.

Instead, the position of phase change interface is observed via thermography, exploiting the variable emission coefficient of the PCM (when the PCM crystallizes,

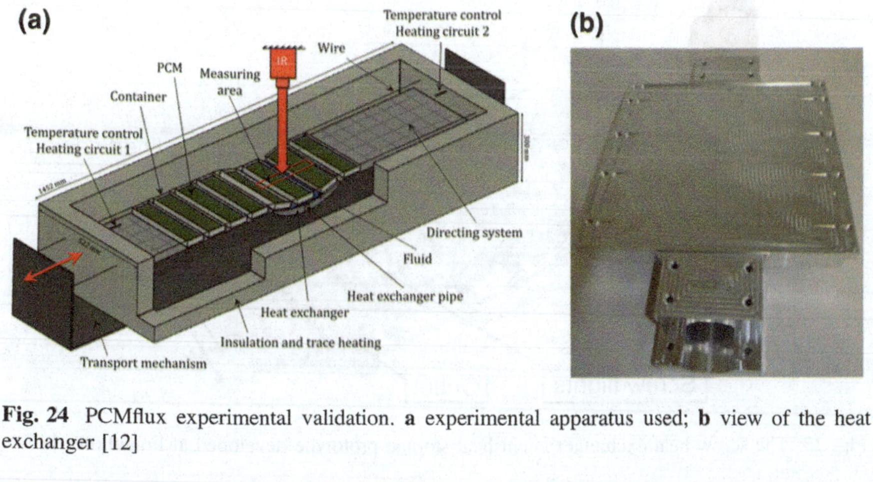

Fig. 24 PCMflux experimental validation. **a** experimental apparatus used; **b** view of the heat exchanger [12]

the emission coefficient increases). In order to properly accomplish such measurements, a ZnSe crystal is inserted in the insulation, covering the inner part of the PCM containers to for the whole length of the heat exchanger to reduce boundary effects.

The experiments were used to verify the operation of PCMflux, but also confirmed the theory behind the proposed concept, since it was possible to locate the phase change interface within the position predicted by a 2D numerical model.

4.3 Dynamic Melting Concept at Fraunhofer ISE: Latent Heat Storage with a Screw Heat Exchanger

In order to address the issue of the low thermal conductivity of PCMs, and respond to the need for a heat storage with sufficient capacity but without too high heat transfer area, a new approach was developed at Fraunhofer ISE: a latent heat storage with a Screw Heat Exchanger (SHE) [13].

SHEs typically have one or more parallel transport screws enclosed in a double lined trough. By turning the screws, material is conveyed from inlet to outlet of the SHE. At the same time, the material can be heated or cooled by a heat transfer medium (water, thermal oil, steam) flowing through ducts in the trough, the shafts and the screw threads.

In the proposed concept, the PCM is transported along the heat exchanger area inside the SHE, while the HTF flows through the hollow shaft, screws. Phase change occurs during the passage in the SHE, but the solid and liquid material is stored elsewhere, i.e. in separated containers, thus decoupling the power needed for the phase change process from the storage capacity of the system. A picture of the

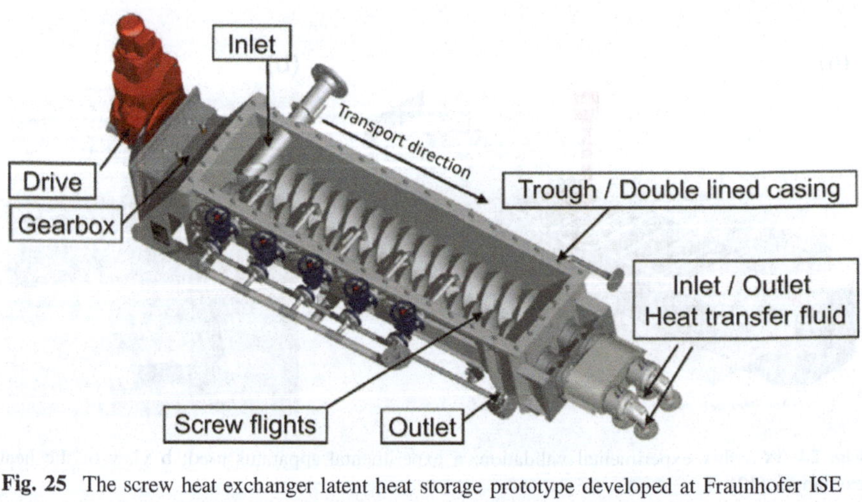

Fig. 25 The screw heat exchanger latent heat storage prototype developed at Fraunhofer ISE

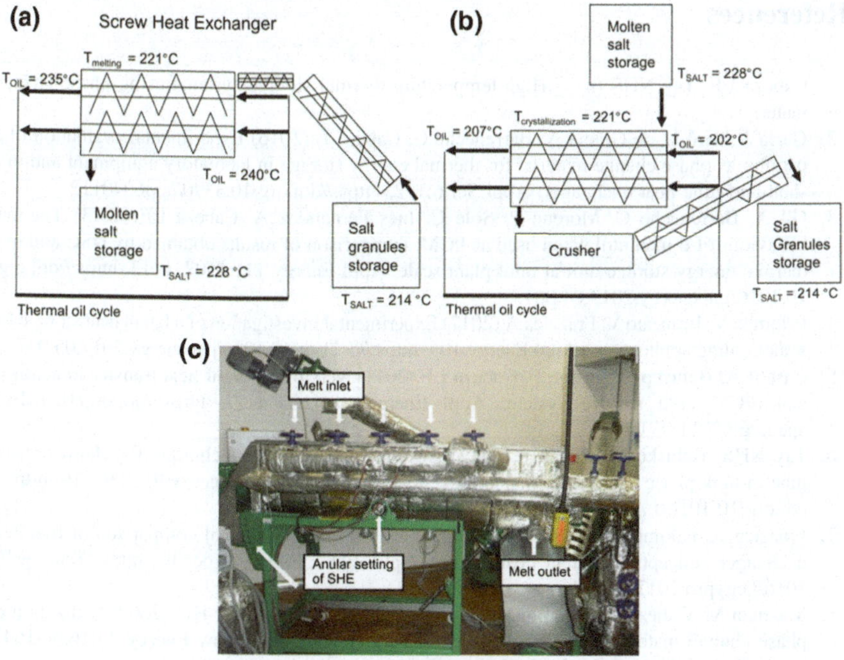

Fig. 26 **a** Layout of the experimental apparatus for the evaluation of the proposed SHE concept during charge; **b** layout of the experimental apparatus during discharge; **c** side view of the apparatus installed in the laboratory at ISE [13]

prototype developed is shown in Fig. 25. The chosen PCM is an eutectic mixture of sodium nitrate and potassium nitrate. The prototype was equipped with RTDs for measuring the temperature of the PCM. Instead, the HTF circuit was equipped with pressure sensors, resistance temperature sensors and a piston rotary.

The testing rig used for the experimental campaign is shown in Fig. 26: apart from the storage for the molten and solid salt, it includes the SHE latent heat storage, connected to a diathermic oil heater. A transport system for the solid and liquid salt is also part of the layout: the molten salt tank is moved with a crane between inlet and outlet, while the solid granular PCM is transported by means of three transport screws. Between outlet and storage tank, a crusher is installed, to allow the adaption of the bulk density of the granules during the experiments. These were aimed at proving the possible successful operation and calculating heat transfer coefficient. Outcomes of the measurements was the continuous phase change from liquid to solid inside the SHE during the transport of the material. The self-cleaning effect in the SHE proved sufficient to clean the heat exchanger surface from growing salt layers and to create granular salt. Limitations in the heat transfer coefficient were mainly found to be on the salt side.

References

1. Cabeza LF, Tay NHS (n.d.) High-temperature thermal storage systems using phase change materials
2. Gasia J, Miró L, de Gracia A, Barreneche C, Cabeza L (2016) Experimental evaluation of a paraffin as phase change material for thermal energy storage in laboratory equipment and in a shell-and-tube heat exchanger. Appl Sci 6:112. https://doi.org/10.3390/app6040112
3. Gil A, Barreneche C, Moreno P, Solé C, Inés Fernández A, Cabeza LF (2013) Thermal behaviour of d-mannitol when used as PCM: comparison of results obtained by DSC and in a thermal energy storage unit at pilot plant scale. Appl Energy 111:1107–1113. https://doi.org/10.1016/j.apenergy.2013.04.081
4. Palomba V, Brancato V, Frazzica A (2017) Experimental investigation of a latent heat storage for solar cooling applications. Appl Energy 199. https://doi.org/10.1016/j.apenergy.2017.05.037
5. Castell A, Belusko M, Bruno F, Cabeza LF (2011) Maximisation of heat transfer in a coil in tank PCM cold storage system. Appl Energy 88:4120–4127. https://doi.org/10.1016/j.apenergy.2011.03.046
6. Tay NHS, Belusko M, Bruno F (2012) An effectiveness-NTU technique for characterising tube-in-tank phase change thermal energy storage systems. Appl Energy 91:309–319. https://doi.org/10.1016/J.APENERGY.2011.09.039
7. Frazzica A, Palomba V, Rosa DL, Brancato V (2017) Experimental comparison of two heat exchanger concepts for latent heat storage applications. Energy Procedia. https://doi.org/10.1016/j.egypro.2017.09.501
8. Mazman M, Cabeza LF, Mehling H, Nogues M, Evliya H, Paksoy HÖ (2009) Utilization of phase change materials in solar domestic hot water systems. Renew Energy 34:1639–1643. https://doi.org/10.1016/J.RENENE.2008.10.016
9. de Gracia A, Oró E, Farid MM, Cabeza LF (2011) Thermal analysis of including phase change material in a domestic hot water cylinder. Appl Therm Eng 31:3938–3945. https://doi.org/10.1016/J.APPLTHERMALENG.2011.07.043
10. Wu J, Tremeac B, Terrier M-F, Charni M, Gagnière E, Couenne F, Hamroun B, Jallut C (2016) Experimental investigation of the dynamic behavior of a large-scale refrigeration—PCM energy storage system. Validation of a complete model. Energy 116:32–42. https://doi.org/10.1016/J.ENERGY.2016.09.098
11. Tay NHS, Bruno F, Belusko M (2013) Experimental investigation of dynamic melting in a tube-in-tank PCM system. Appl Energy 104:137–148. https://doi.org/10.1016/J.APENERGY.2012.11.035
12. Pointner H, Steinmann W-D (2016) Experimental demonstration of an active latent heat storage concept. Appl Energy 168:661–671. https://doi.org/10.1016/J.APENERGY.2016.01.113
13. Zipf V, Neuhäuser A, Willert D, Nitz P, Gschwander S, Platzer W (2013) High temperature latent heat storage with a screw heat exchanger: design of prototype. Appl Energy 109:462–469. https://doi.org/10.1016/J.APENERGY.2012.11.044

Experimental Characterization of Sorption Thermal Energy Storage Systems

Valeria Palomba, Salvatore Vasta and Andrea Frazzica

Abstract In the present chapter, the experimental methods employed for the characterisation of sorption TES are discussed. The investigated systems comprise liquid and solid sorption technologies, both for closed and open systems. In particular, the proposed procedure for the closed sorption TES systems is described in details and an example on a lab-scale adsorption TES is reported. Finally, also experimental methodologies applied for components testing, namely, kinetic testing of small-scale adsorbers and characterisation of evaporators, are introduced.

1 Introduction

Physical sorption technology has been the centre of an intense research for many years, even though the focus used to be identified with space cooling and heating applications. Recently, energy storage through physical sorption has gained interest, profiting also from the experience in the field of absorption and adsorption for air conditioning.

Within the present chapter, an overview of experimental methodologies successfully applied for the evaluation of system performance is given. Both liquid absorption and solid adsorption are considered. For the latter case, the approach in the description of experimental cases and apparatuses is twofold: first of all, valuation methods for open and closed systems, presenting different features and peculiarities will be presented. Subsequently, valuation methods for open and closed systems, presenting different features and peculiarities will be presented.

V. Palomba (✉) · S. Vasta · A. Frazzica
CNR-ITAE, Istituto di Tecnologie Avanzate per l'Energia "Nicola Giordano",
Via Salita S. Lucia sopra Contesse 5, 98126 Santa Lucia, Messina, Italy
e-mail: valeria.palomba@itae.cnr.it

A. Frazzica and L. F. Cabeza (eds.), *Recent Advancements in Materials and Systems for Thermal Energy Storage*, Green Energy and Technology,
https://doi.org/10.1007/978-3-319-96640-3_12

2 Classification of Experimental Systems

The most general classification of experimental systems in the physical sorption TES field, which is shown in Fig. 1, is between the two core technologies, i.e. liquid absorption and solid adsorption. Within solid adsorption technology, that is experiencing a significant boost, open and closed systems can be distinguished: while the former ones make use of air as heat transfer medium, the latter systems are designed to work with heat transfer fluids in closed loops. In both cases, the main application of the technology for storage purposes is the integration with space heating/cooling system, often coupled with renewables.

3 Experimental Methods with Liquid Absorption: The $NaOH/H_2O$ TES Concept

Liquid absorption storage technology with the working pair NaOH/water is the leading research focus realised by SPF Institute of HSR University for Applied Sciences in Rapperswill and EMPA. Experimental methods employed for the characterization of systems include both the evaluation of the core components of the storage, i.e. the heat exchangers employed, and the operational campaign on a prototype [1–3].

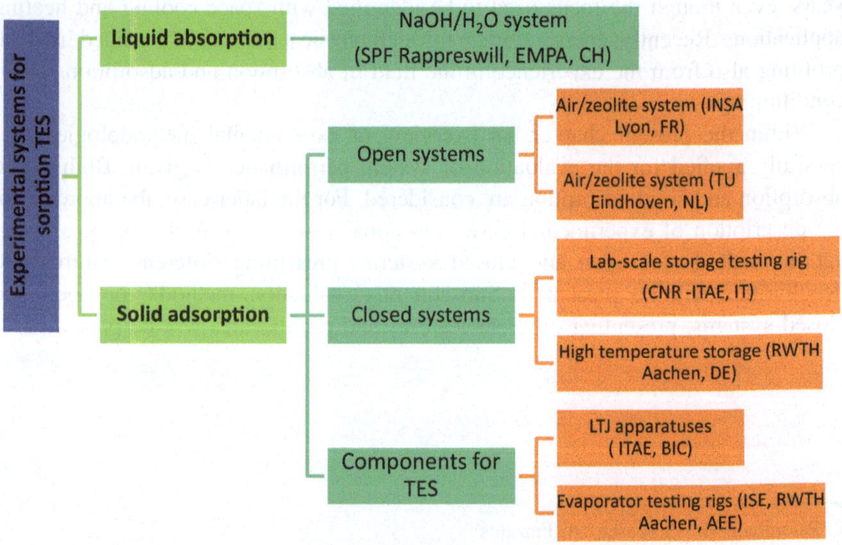

Fig. 1 Classification of the experimental systems covered in the chapter

Fig. 2 Pictures of the TES pilot system with NaOH/water [1]

The complete pilot system, shown in Fig. 2, is described in [1]. It was built into a shipping container. The solar field, which represents the heat source for the operation, is installed on the outside of the container. The 8 kW sorption storage consists of two tube bundle falling film heat and mass exchangers, one serving as absorber and desorber and the other operating as condenser and evaporator. They are sideways interconnected through a large opening for unhindered vapour exchange. The demonstrator contains also three stainless steel storage vessels of 1.5 m^3 each and gear pumps for the transport the sorbent (aqueous NaOH) and sorbate (water) to and from the heat and mass exchangers.

In order to test the absorption process the demonstrator was filled with 1000 kg of sodium hydroxide at a concentration of 50 wt% NaOH in water and 700 kg of de-ionised water. In order to dissolve the gases in the solution, the vacuum pump is always operated prior to testing. The testing is reported for the case of desorption. In particular, temperatures in the heat transfer fluid circuit and in several points of the heat and mass exchangers were measured and reported. The preliminary results presented indicate that a power output of 1 kW was achieved. More importantly, the importance of the correct design of HEXs was discussed: the heat and mass transfer exchangers are tube bundles usually employed in absorption chillers, but they significantly limit the performance of the system, due to the short time that the sorbent remains in the sorbate atmosphere.

Such an experimental facility was subsequently employed for testing a new design of the heat and mass exchangers for the sorption unity. A CAD view of these systems, presented in [2], is shown in Fig. 3. For its experimental testing, the following sensors were employed:

- Pt100 temperature sensors for the measurements of temperature in the HTF circuits and the outside temperature in the tube bundles;
- piezoelectric pressure sensors for the measurement of pressure in the vacuum vessels;
- vortex flow metres for measuring flow rates in all the circuits inside of the tubes;
- dosing pumps for measuring the levels on the outside of the tubes;

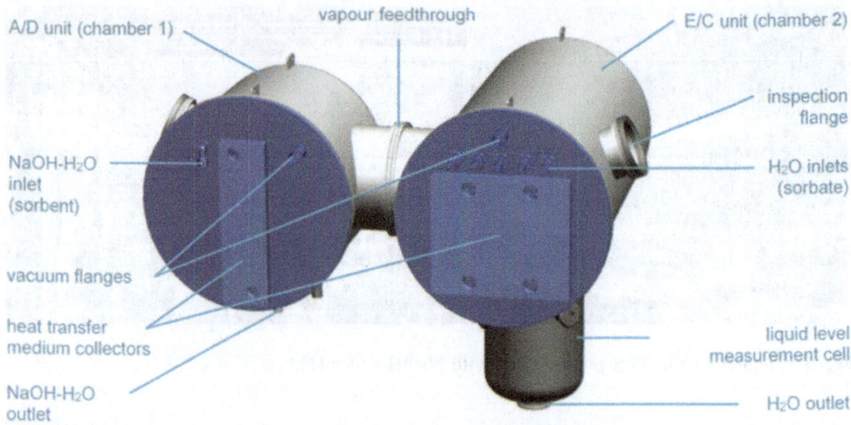

Fig. 3 CAD view of the new design for the NaOH/water TES [2]

- a conductivity metre to measure the concentration of the solution in the absorber/desorber unit.

All the signals are acquired and recorded using a NI® system and LabVIEW® interface. The selected time step was 3.0 s, but each test was analysed averaging data of a 30 min measurement time scale, that enables to overcome small signal instabilities due to flow fluctuations on the absorber/desorber unit.

The aim of the experimental campaign was to characterise the absorber/desorber unit and therefore is was considered necessary to run the evaporator/condenser in a way that it does not limit the vapour exchanges, i.e. selecting high heat transfer fluid flow rates. The drawback of such a choice is that a higher uncertainty affects power measurement on the evaporator/condenser side.

The results of the experimental campaign were presented in terms of heat transfer coefficients and powers, as a function of the temperature differences between absorber and evaporator, and mass flow rate of the working fluid. The main outcome was that, even though the system could be successfully operated, further improvement is needed, in particular to increase absorption rate, through a more efficient heat exchanger. From the measurement point of view, it was pointed out that the choice of increasing the flow rate at the evaporator/condenser does not allow determining the exact amount of water evaporated/condensed. Moreover, the time response of the concentration sensor used was not sufficient to give information on instant concentration.

Since heat exchanger design proved to be a key issue in the correct operation of the storage, a dedicated setup was built for the evaluation of a new type of heat and mass exchanger apparatus, which is described in [3]. A picture of the whole system is shown in Fig. 4. It is mainly composed of two heat exchangers to be tested, that are spiral fin-tube heat exchangers vertically arranged. They are enclosed into

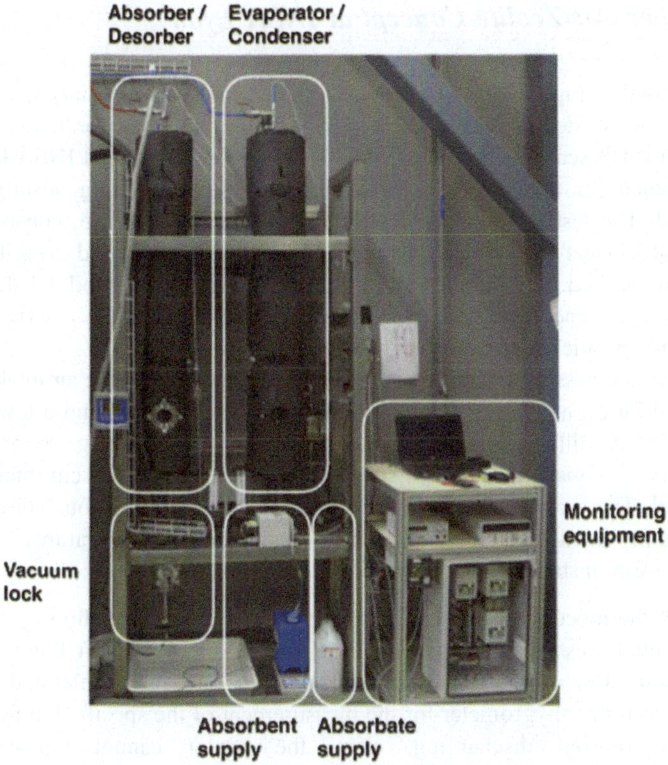

Fig. 4 Experimental apparatus for the evaluation of heat exchangers for absorption TES [3]

vacuum vessels that are connected by means of two vapour exchange channels. Heat sources/sinks are simulated by means of thermostatic baths. Gear pumps are used to circulate the HTF, whereas dosing pumps are used for the working fluid. A glass port is installed to allow confirming absorbate wetting of the heat exchanger. Moreover, a vacuum lock is installed, to allow periodic release and sampling of the absorbent concentration without process interruption. Regarding testing procedure, both absorption and desorption were tested, changing different parameters alongside the tests. That is, temperature lift, mass flow rates and evaporation and condensation temperatures. With respect to one reference test for absorption and one for desorption, all the other tests were performed changing one parameter at time.

Results of measurements highlighted that high concentration difference between absorption and desorption or charged and discharged state is the key parameter for efficient operation. In order to achieve this target, important parameters are large surface area, good wetting, thin film and sufficient exposure time. Results also indicated that there is a trade-off between energy storage capacity and specific power output.

3.1 *Open Air/Zeolite Concept at INSA Lyon*

Zeolite-based storages have been proposed during the years to satisfy space heating needs in the residential sector. However, most of the experimentation was performed in small-scale or lab-scale levels. To cover for this gap, at INSA Lyon, one testing bench for the testing of zeolite/air heat storages using adsorbents was developed. The test bench, described in [4] and shown in Fig. 5, consists of two cylindrical vessels of 140 l, containing zeolite beads, disposed over the entire section of the bed. A peculiarity of the testing bench is its modular design that permits to experiment different configurations: 'one vessel alone', 'series configuration' and 'parallel configuration'.

Each zeolite vessel consists of an upper region dedicated to the air intake, thanks to an air diffuser, a central region containing the zeolite beads and a lower region with another air diffuser for the homogeneous outlet of the airflow. The vessels and the air pipes are made of stainless steel and are insulated by a 5 cm thick layer of glass wool. The sealing at the contact surfaces between the various zones of each module is provided by a silicone gasket resistant to high temperatures.

The sensors installed include

- type *K* thermocouples to measure the temperature of the air flow;
- differential pressure sensors to measure the pressure of the air flow;
- ultrasonic flow metres to measure volume flow rates at the inlet and outlet;
- a chilled mirror hygrometer for the measurement of the specific humidity of the airflow (during discharging, since the sensor cannot withstand high temperatures).

Time step selected for the acquisition of recorded parameters is 30 s.

Two different charge temperatures (120 and 180 °C) were tested, while discharge temperature was set to 20 °C with 70% RH of inlet air. Results showed that an average temperature lift of 38 °C per vessel could be obtained, for a total duration of discharge of 8 h. These results were used to propose a 1D simplified model of the reactors.

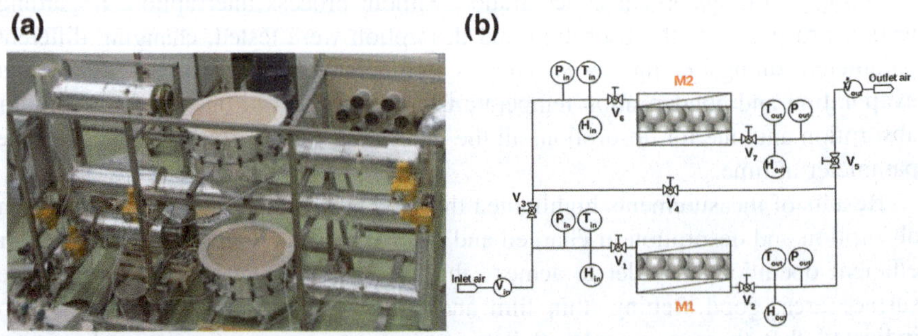

Fig. 5 The testing bench for the evaluation of zeolite/air TES at INSA Lyon. **a** Picture of the system; **b** schematics [4]

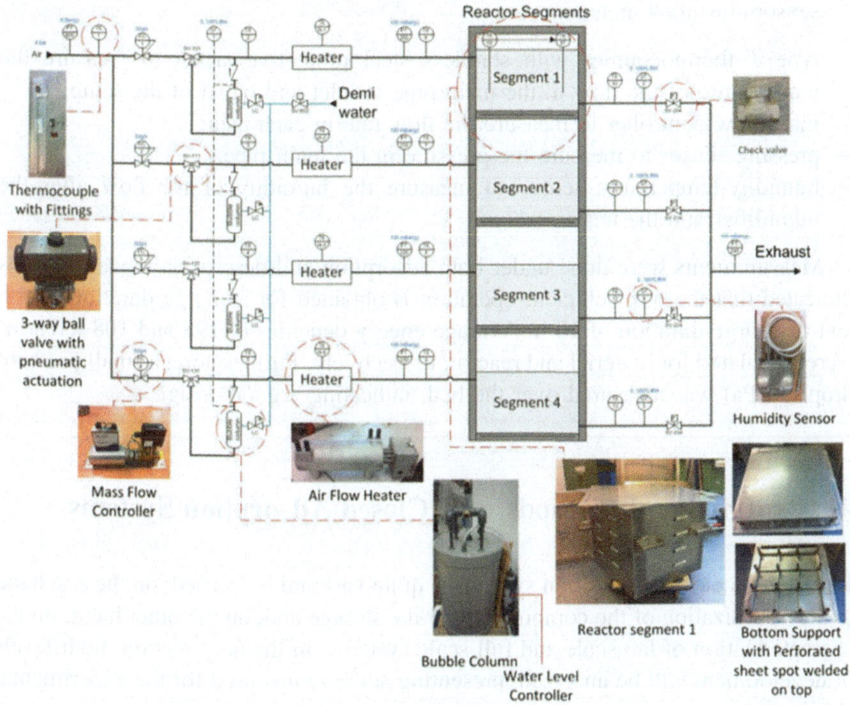

Fig. 6 Pilot plant to evaluate open adsorption TES at TU Eindhoven [5]

3.2 *Open air/zeolite concept at TU Eindhoven*

A pilot setup for the evaluation of open adsorption storages was developed at Eindhoven University of Technology, consisting of four segments. It was designed with a modular feature, in order to be able to operate under series or parallel configuration, realising up to 4 kW combined power output for at least 10 h. The setup is described in [5], while the schematics, including pictures of the main components, are presented in Fig. 6.

In order to have flexibility in the operation of multiple segments, each segment is connected to an individual pipeline. Air flows into the system from the main inlet pipe, and then is divided over four paths. In each path, the air flow can be conditioned for either charging (dehydration) or discharging (hydration); during the hydration process the flow is humidified in a bubble column. Afterwards, the flow passes through a heater, used to heat up the dry air during the dehydration process or to preheat the humid air during the hydration process. The outlet of each reactor segment is connected to a check valve and then to the exhaust vent. The purpose of these check valves is to close the outlet when there is no flow after the experiments, to prevent any moisture being exchanged between the surroundings and the reactor segments.

Sensors installed include:

- type T thermocouples with stainless steel protective jacket to measure the temperature of the flow in the main pipe at inlet and outlet of the setup;
- mass flow controller to measure the flow rate in each path;
- pressure sensor to measure the pressure in the main pipe;
- humidity-temperature sensor to measure the humidity of the flow after the humidifier and the heater.

Measurements were done under both adsorption and desorption modes. Results indicated that the most efficient operation is obtained for a charge duration of 5 h and discharge duration of 10 h. Average energy densities of 198 and 108 kWh/m^3 were calculated for material and reactor, respectively. Furthermore, a small pressure drop (30 Pa) was measured over the bed, indicating a good design.

4 Experimental Methods with Closed Adsorption Systems

Research on closed adsorption systems is quite vast and is focused, on the one hand to the optimization of the components for the storage and, on the other hand, on the characterization of lab-scale and full-scale systems. In the next section, both levels of development will be analysed, presenting some setups used for the experimental characterization.

4.1 High Temperature TES Unit at RWTH Aachen

An energy storage unit for evaluation of zeolite 13X/water system was built at RWTH Aachen. The system is presented in [6], while a picture of the system is shown in Fig. 7.

The system mainly consists of two vacuum vessels, an adsorber module and an evaporator/condenser. They are connected by means of a vacuum valve. The stainless steel adsorber heat exchanger is made of a bundle of tubes connected by lamellae and is filled with zeolite 13X beads. The evaporator contains a stainless steel corrugated-pipe heat exchanger and 2.8 kg of water for adsorption. The vessels are insulated by means of 5 cm of rock wool insulation, that are applied to the outer surface of the adsorber casing and the evaporator. Temperature levels in the adsorber for charging and discharging are ensured by the connection to the circulation system of a thermal oil. During charging, the oil in the circulation system is heated up by an electric heater. During discharging, heat is released via a secondary heat exchanger. The evaporator is heated and cooled via a water circulation system.

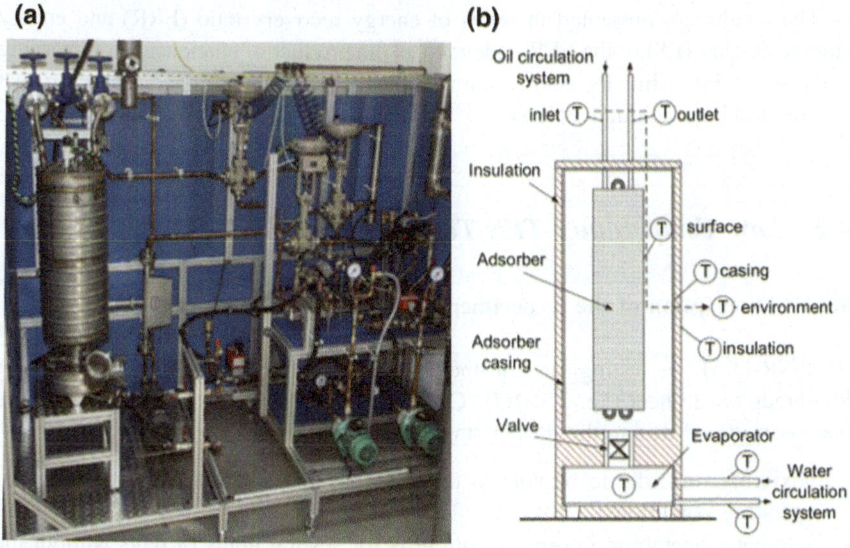

Fig. 7 The experimental setup for high temperature closed adsorption TES at RWTH Aachen. **a** Picture of the system; **b** schematics

Installed sensors include:

- Pt100 for measuring the temperature in the HTFs circuit, the adsorber casing, the insulation and inside the evaporator;
- type *K* thermocouples to measure the ambient temperature and the temperature on the adsorber surface;
- a Coriolis flow meter, for the measurement of the mass flow rate in the oil circulation system.

The system was used to calculate, not only the amount of heat stored, but also to make some detailed calculations about heat losses to the environment. In particular, all measurements to quantify heat losses were performed in steady-state conditions. In particular, temperatures of the adsorber surface and the casing were used for the estimation of heat losses due to convection and radiation.

Subsequently, measures on storage cycles were carried out, either considering a storage period between charging and discharging or under continuous cyclic operation. During storage cycle measurement the procedure is as follows: the system is charged, until the temperature of the adsorber outlet reaches the maximum desorption temperature. After the desired storage time is passed, discharging is induced by cooling of the heat transfer oil and is terminated when the outlet temperature from the adsorber is 70 °C. For each temperature set, at least three successive cycles were measured and the evaluation is started with the second successive cycle to ensure reproducible starting conditions. Instead, during continuous cyclic operation, charging is directly followed by discharging immediately.

The results are presented in terms of energy recovery ratio (ERR) and energy storage density (ESD): the EER is defined as the product of efficiencies for charging and discharging, while the energy storage density is the ratio of the discharging heat and the total storage unit volume.

4.2 *Low-Temperature TES Testing Rig at CNR-ITAE*

4.2.1 Description of the Experimental Setup

At CNR-ITAE, a testing rig for the testing of different sorption storages for low-grade waste heat (i.e. $T < 100$ °C) was developed. The following constraints were considered during the design stage:

- flexibility, in order to be able to test different systems under a wide range of operating conditions;
- automatic operation, in order to run tests for several hours or days without the need for an operator;
- high stability of the thermal levels of all the energy flows provided to the device under test;
- real-time monitoring and recording of all the relevant parameters.

The testing rig consists of two vessels, corresponding to the adsorber chamber for the storage to be tested and an evaporator/condenser (phase changer), connected by means of a gate vacuum valve. The chamber for the adsorber was realised with a flanged removable lid, in order to be able to test different materials and configurations, thus complying with the flexibility demand. The phase changer is realised using four fin-and-tube heat exchangers with copper fins and stainless steel tubes, connected in parallel. The chambers of the adsorber and phase changer are both realised in stainless steel AISI 316L.

The testing rig is completed by a skid, equipped with the hydraulic elements (e.g. electrovalves) and mechanical elements (e.g. pumps and circulators) necessary for the test of the storage unit and the sources/sinks needed to set the temperature levels that simulate the operation of the storage. They consist of an electric heater with power of 36 kW, connected to a 200 l storage tank, to be used as a heat source and a 63 kW electric chiller connected to a 300 l tank, which works as a heat source/sink for the phase changer, thus making available energy flows at the right temperature levels for evaporation and condensation. The desired level of temperature is obtained, both for condenser and evaporator, by means of a thermo-regulator and a mixing valve which mixes the flow in inlet and outlet pipes of the circuit: this method complies with the high-stability temperature flow constraint listed above.

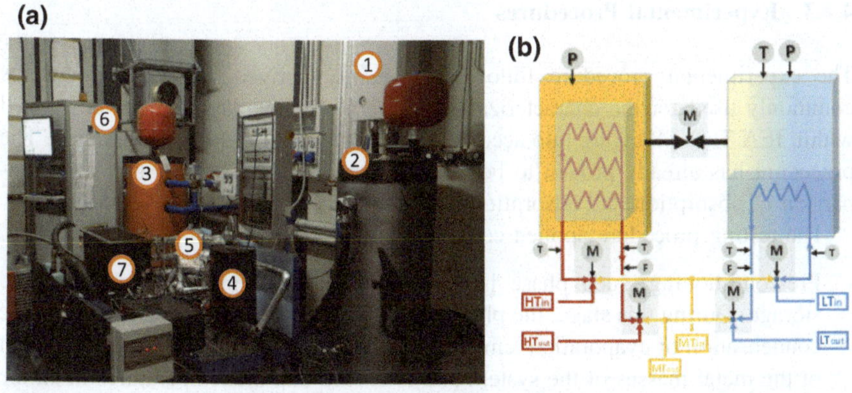

Fig. 8 Testing rig for low-temperature adsorption TES at CNR-ITAE. **a** Picture of the testing bench; 1. heater, 2. 200 l hot storage, 3. 300 l tank, 4. HT buffer; 5. hydraulics; 6. rack for electronics; 7. TES unit. **b** schematic of the TES unit

Finally, a rack contains the electronics, the elements of the acquisition system and a computer for monitoring and recording of all the parameters. The testing rig is described in detail in [7], while its picture and schematics of the skid with the TES unit (i.e. without connections to heat source/sinks) are reported in Fig. 8. All the components were thermally insulated to avoid energy losses with 1.5 cm thick polyurethane foam.

4.2.2 Sensors Employed and Data Acquisition

The management of both the setup for thermal storages and the testing bench is realised through a NI FieldPoint® acquisition system and a dedicated LabVIEW® software. All the components are electrically operated, to fulfil the requirements of real-time acquisition and automatic operation.

The following sensors are installed:

- Pt 100 1/10 DIN with four wires, used for the measurement of inlet and outlet temperatures of the HTF both at the adsorber and the evaporator/condenser;
- Class A *T*-type thermocouple for the measurement of the temperature of the refrigerant inside the phase changer and the temperature of the adsorbent material;
- magnetic flow meter with 1% FS accuracy to measure the flow rate of the HTF in the adsorber and the evaporator/condenser;
- piezoresistive pressure sensors with 0.25% FS accuracy to measure the pressure inside both vacuum chambers.

4.2.3 Experimental Procedures

The experimental procedure followed is mainly derived from the procedures commonly used for the characterization of thermally operated chillers, developed within IEA Task 48 taking into account the specific features of TES operation. The procedure has already proven to be a useful method for the evaluation and comparison of absorption and adsorption systems [8].

The testing procedure applied consists of three main steps:

- Pre-conditioning: such phase is realised before charging or discharging the storage. During this stage, the phase changer is heated up or cooled down at the condensation or evaporation temperature, in order to minimise the inertial effect of the metal masses of the system and to ensure that the temperatures in all the components actually match the set points. Pre-conditioning is stopped once steady-state conditions are reached, i.e. when the power output from the phase changer is around the uncertainty of measurements.
- Charge: it is realised by heating up the adsorber and opening the connection to the condenser. Condensation occurs at the phase changer and the condensation heat is removed.
- Discharge: it is realised by cooling down the adsorber and opening the connection to the evaporator. Evaporation occurs at the phase changer and the evaporation heat is provided.

Similarly to the case of the test bench at RWTH in Aachen, at least three successive cycles were measured and the evaluation is started with the second successive cycle to ensure reproducible starting conditions.

The following parameters were calculated:

- Discharging Power (DP), defined as the average power provided by the system during discharging period.
- Stored Energy (SE), defined as the total amount of energy supplied to the user.
- Storage Energy Density (SED), defined as the specific energy per unit mass of the material.
- Storage efficiency (η), defined as the ratio of the stored energy (which represents the energy actually recovered during one cycle) over the heat input supplied to the system.

For each tested configuration, three possible operating modes are possible: cold storage, short-term heat storage and long-term heat storage. The equations used for the calculations of the performance parameters are summarised in Table 1.

4.2.4 Example of System Testing

In the following, the testing method and the results obtained by applying the procedure above described, on a simple sorption storage system, are reported.

Table 1 Parameters calculated for the experimental characterization of low-temperature adsorption TES at CNR-ITAE

Cold storage	
Discharge power (DP)	$PD = \frac{\int_0^{t_{ev}} \dot{m}_{pc} c_p (T_{in,pc} - T_{out,pc}) dt}{t_{ev}}$ (W)
Stored energy (SE)	$SE = \int_0^{t_{ev}} \dot{m}_{pc} c_p (T_{in,pc} - T_{out,pc}) dt$ (Wh)
Storage energy density (SED)	$SED = \frac{SE}{m_{adsorbent}}$ (Wh/kg)
Storage efficiency (η)	$\eta = \frac{\int_0^{t_{ev}} \dot{m}_{pc} c_p (T_{in,pc} - T_{out,pc}) dt}{\int_0^{t_{des}} \dot{m}_{ads} c_p (T_{in,ads} - T_{out,ads}) dt}$
Long-term heat storage	
Discharge power (DP)	$PD = \frac{\int_0^{t_{adsorption}} \dot{m}_{ads} c_p (T_{out,ads} - T_{in,ads}) dt}{t_{adsorption}}$ (W)
Stored energy (SE)	$SE = \int_0^{t_{adsorption}} \dot{m}_{ads} c_p (T_{out,ads} - T_{in,ads}) dt$ (Wh)
Storage energy density (SED)	$SED = \frac{SE}{m_{adsorbent}}$ (Wh/kg)
Storage efficiency (η)	$\eta = \frac{\int_0^{t_{adsorption}} \dot{m}_{ads} c_p (T_{out,ads} - T_{in,ads}) dt}{\int_0^{t_{des}} \dot{m}_{ads} c_p (T_{out,ads} - T_{in,ads}) dt}$
Short-term heat storage	
Discharge power (DP)	$PD = \frac{\int_0^{t_{cond}} \dot{m}_{pc} c_p (T_{out,pc} - T_{in,pc}) dt + \int_0^{t_{adsorption}} \dot{m}_{ads} c_p (T_{out,ads} - T_{in,ads}) dt}{t_{cond} + t_{adsorption}}$ (W)
Stored energy (SE)	$SE = \int_0^{t_{cond}} \dot{m}_{pc} c_p (T_{out,pc} - T_{in,pc}) dt + \int_0^{t_{adsorption}} \dot{m}_{ads} c_p (T_{out,ads} - T_{in,ads}) dt$ (Wh)
Storage energy density (SED)	$SED = \frac{SE}{m_{adsorbent}}$ (Wh/kg)
Storage efficiency (η)	$PD = \frac{\int_0^{t_{cond}} \dot{m}_{pc} c_p (T_{out,pc} - T_{in,pc}) dt + \int_0^{t_{adsorption}} \dot{m}_{ads} c_p (T_{out,ads} - T_{in,ads}) dt}{\int_0^{t_{des}} \dot{m}_{ads} c_p (T_{out,ads} - T_{in,ads}) dt}$

A first experimental campaign was realised employing, for the adsorber, a flat-tube and fins heat exchanger with exchange area of 1.75 m^2, that was filled with 4.3 kg of AQSOA FAM Z02 grains, in the range 1–2 mm. The material was contained inside the heat exchanger by means of a metallic net.

4.2.5 Typical Results

Figure 9 shows the temperatures measured during a test using the setup described above, where the different phases are highlighted. It is possible to notice that the oscillations in temperatures are very low, thus confirming the reliability of the

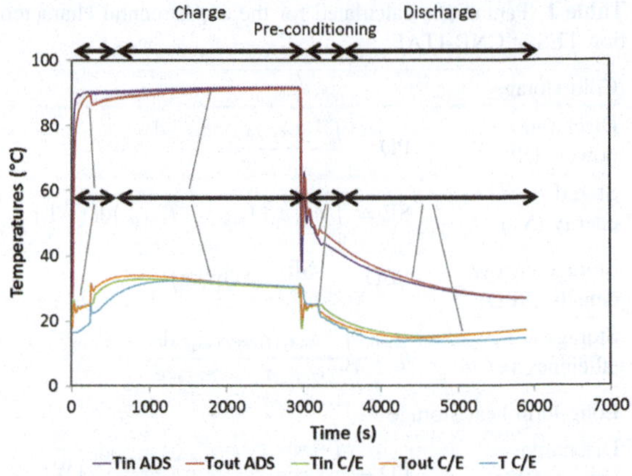

Fig. 9 Temperatures recorded during a test in the testing rig at CNR-ITAE

experimental setup and of the procedure applied. Testing conditions are $T_{des} = 90$ °C, $T_{adsorption} = 35$ °C, $= T_{cond} = 35$ °C, $T_{ev} = 5$ °C, $\dot{m}_{PC} = 10\,\text{kg/min}$, $\dot{m}_{Ads} = 12\,\text{kg/min}$. A fixed time of 55 min for charge was chosen. During charge, the adsorber is heated up to 90 °C, at closed volume, while the phase changer that acts as a condenser, is thermally regulated to the set temperature. After a short conditioning time, the connection between the adsorber and the condenser is opened and the water vapour is desorbed and condensed and stored inside the phase changer. The temperature of the adsorber is almost constant throughout the whole process, while the temperature of the condenser experiences an initial increase of about 5 °C, due to the beginning of condensation and then it remains stable at 35 °C. Afterwards, a short conditioning phase at closed volume is realised, during which the liquid in the phase changer is cooled to the set evaporation temperature and the adsorber is cooled down: this phase simulates an undefined time storage period. Subsequently, the connection between the adsorber and the evaporator is opened again. This allows the evaporation of the refrigerant that is adsorbed onto the zeolite. The temperature inside the adsorber experiences an initial peak and subsequently decreases until the measured power approaches zero, and the test is therefore stopped.

Figure 10 shows the powers measured during the same test. During both charge and discharge, the measured power presents an initial peak and subsequently decreases. The maximum charging power measured is around 3.5 kW, while during discharge a peak power of around 1.5 kW was measured.

4.2.6 Influence of Operating Conditions

The complete analysis of the results and operation features is given in [7]. Here, as an exemplary case of the measured values and possible analysis of the effect of

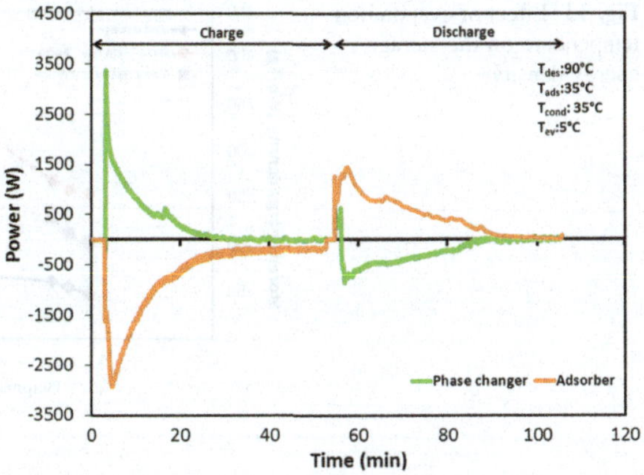

Fig. 10 Powers record during a test in the testing rig at CNR-ITAE

some process parameters is discussed. The experimental campaign carried out was aimed at defining the effect of operating conditions. To do so, different desorption temperatures, condensation temperatures and evaporation temperatures were tested. Figures 11 and 12 show the effect of desorption and evaporation temperature, respectively, on the storage energy density in the three operating modes examined. Desorption temperature holds a significant effect on the energy actually recoverable: in a short-term heat storage operation, passing from 75 to 90 °C corresponds to a 60% increase in the recovered energy. Similarly, for long-term heat storage, going from 75 to 90 °C allows increasing the storage density from 100 to 150 Wh/kg. If the storage density is calculated on a volumetric basis (considering the volume of the adsorber only), the measured values corresponds to energy densities of 60 up to 120 kWh/m^3.

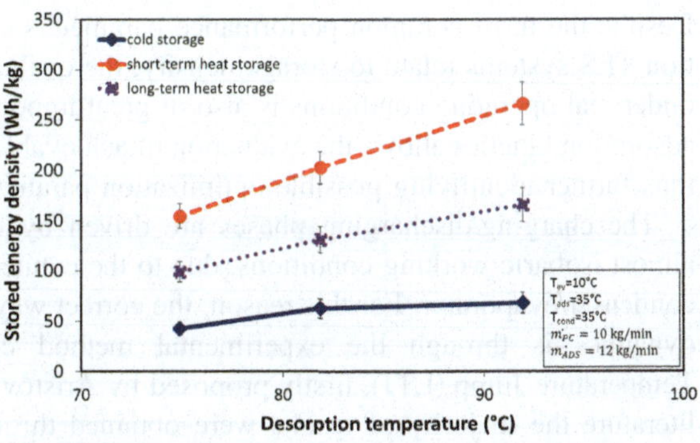

Fig. 11 Effect of desorption temperature on the storage energy density

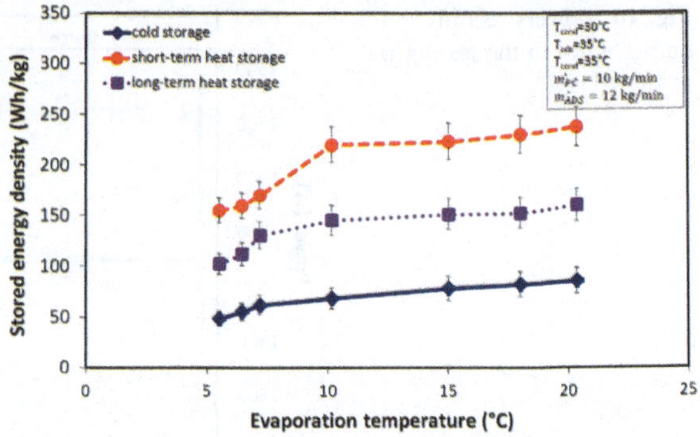

Fig. 12 Effect of evaporation temperature on the storage energy density

The effect of evaporation temperature is less marked, with the exception of cold storage operation. Indeed, for cold storage operation, the energy recoverable with the increasing evaporation temperatures from 5 to 20 °C passes from 50 to 100 Wh/kg. The effect on heat storage operation is more evident only in the range 5–10 °C, while a plateau is reached afterwards. On a volumetric basis, such values correspond to energy densities in the range 50–120 kWh/m^3, which is almost doubled with respect to the energy density of water under the same boundary conditions.

5 Experimental Methods for Characterization of Components of Adsorption Systems

5.1 LTJ Apparatuses

Despite the most common performance parameters commonly evaluated for sorption TES systems relate to storage density, the evaluation of the sorption dynamic under real operating conditions is also of great importance. Indeed, the desorption/adsorption kinetics allows the evaluation of achievable charging/discharging power, thus further identifying possible optimization parameters for the TES design.

The charging/discharging phases are driven by temperature jump/drop under almost isobaric working conditions, due to the connection of the adsorber with the condenser/evaporator. For this reason, the correct way to evaluate the sorption TES dynamics is through the experimental method commonly defined as Large Temperature Jump (LTJ), firstly proposed by Aristov et al. in [9]. Currently, in the literature the majority of results were obtained through two versions of the LTJ approach, the volumetric one, developed at the Boreskov Institute of Catalysis (BIC), and the gravimetric one, developed at the CNR-ITAE.

Figure 13 reports the schematic of the volumetric LTJ apparatus developed and realised at BIC. Since it is based on volumetric approach, the sorption kinetic is indirectly measured by evaluating the pressure evolution inside a closed known volume. Indeed, the sorbent material, placed on a flat plate heat exchange in the measuring cell is firstly equilibrated at a certain temperature and vapour pressure, see points 2 and 4 in Fig. 14. The equilibrium is achieved by keeping constant the temperature of the plate heat exchanger and connecting the measuring cell to a big

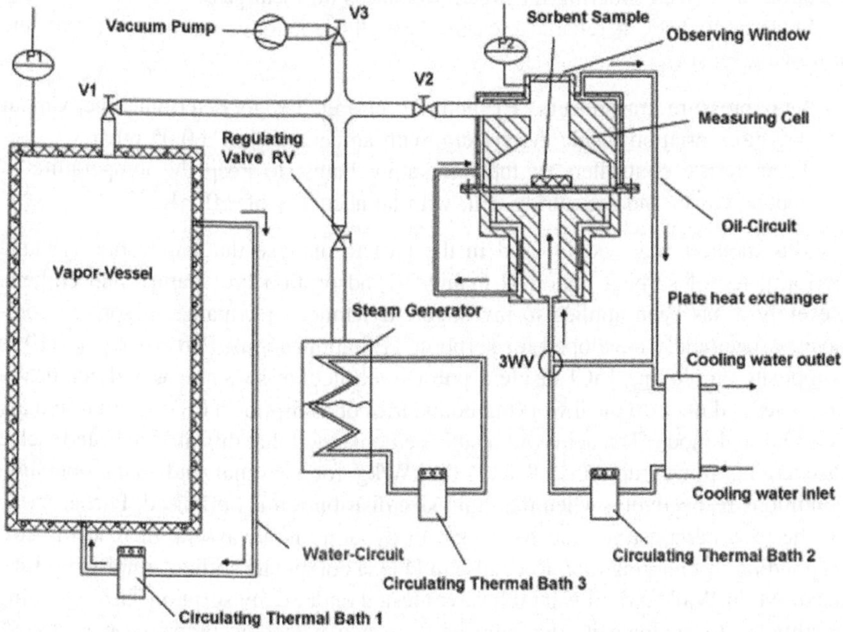

Fig. 13 Schematic of the volumetric LTJ apparatus available at the Boreskov Institute of Catalysis [9]

Fig. 14 Quasi-isobaric desorption and adsorption phases on an isosteric chart [9]

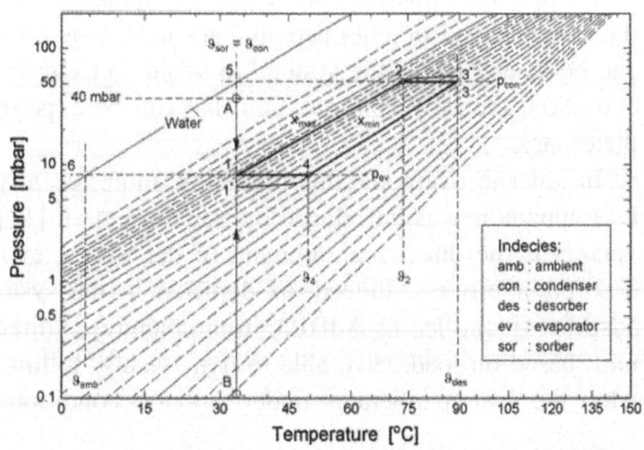

vapour vessel imposing the saturation vapour pressure. Subsequently, the temperature of the sample is suddenly jumped up/dropped down, causing a desorption/adsorption which is registered by a pressure variation inside the apparatus. Since the amount of adsorbent material tested is limited (i.e. about 100 mg) and the volume of the system is big enough, the total pressure variation is kept within 2–3 mbar, which allows defining this measurement as quasi-isobaric. The equilibrium ending point 1′ and 3′, in Fig. 14, slightly differ from the theoretical ones, i.e. points 1 and 3.

The sorption kinetic is then evaluated by converting the pressure evolution in sorption rate by considering the water vapour as an ideal gas.

In order to have a reliable measurement, the apparatus is equipped with the following sensors:

- Two pressure transducers, P1 and P2 in Fig. 13, to determine the vapour pressure variation inside the system, with an accuracy of ±0.05 mbar.
- Temperature controllers of the circulating baths, to keep the temperature of vapour vessel and measuring cell, with an accuracy of ±0.1 K.

This method was widely used in the past to analyse the achievable dynamic performance of sorbent materials mainly for adsorption heat pumps and chillers. Recently it has been applied to investigate dynamic performance of sorbent composites specifically developed for sorption TES applications. Particularly, in [10] a composite employing LiCl inside a porous vermiculite structure was dynamically characterized under typical working boundaries of a sorption TES operating in daily and seasonal mode. The achieved results demonstrated that this material can reach a discharging power up to 11.8 and 3.9 kW/kg for seasonal and daily operating conditions respectively, when the grain size distribution is optimised. Furthermore, all the tests demonstrate that the charging dynamic is always faster than the corresponding discharging one. Recently, in [11], a composite sorbent employing LiCl inside Multi Wall Carbon Nanotubes was tested under daily sorption TES operating conditions, to evaluate its dynamic performance when the preparation is varied, from grains, to pellets obtained by pressing and pellets formed with a binder. The main achieved results confirmed that employing a binder to prepare pellets allows increasing a lot the dynamic performance, with highest value of 11.2 kW/kg.

Despite the attractive features of this apparatus, the main limitation is related to the amount of adsorbent material that can be tested. Indeed, since only 100–200 mg can be loaded inside the system, to avoid an excessive deviation from the isobaric evolution, the only configuration that can be experimentally replicated is the flat plate one.

In order to overcome this limitation, more recently a gravimetric version of the LTJ apparatus was introduced by Sapienza et al. [12]. The main difference of this version is the direct measurement of the weight evolution during the adsorption/desorption process. Indeed, as depicted in the system schematic in Fig. 15, the adsorber (identified as AdHEX in the picture) is directly connected to a weighting unit, based on load cells, able to dynamically follow the weight evolution. In this way, the system operates under isobaric conditions, since there is no need of

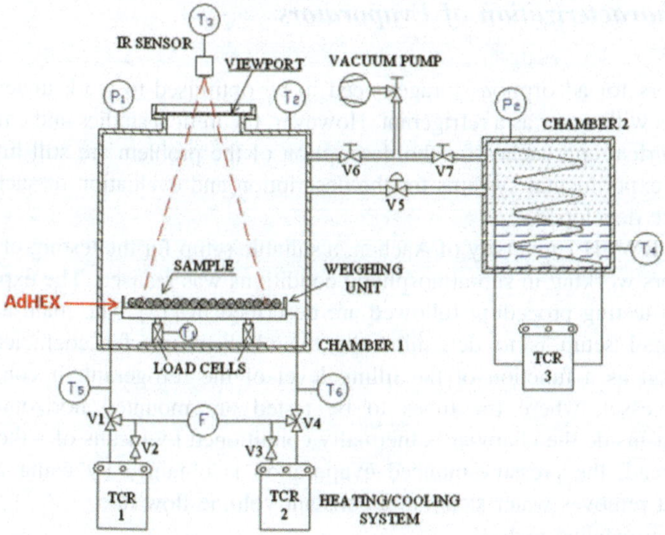

Fig. 15 Schematic of the gravimetric LTJ apparatus at the CNR-ITAE [12]

measuring pressure variation, and it is able to measure the dynamics of different adsorber configurations, since the amount of adsorbent material that can be tested only depends on the features of the installed load cells. Indeed, different adsorber configurations as well as working pairs were tested so far by means of this apparatus [13–16].

The system operation follows the one already defined for the volumetric apparatus. In this case, the measurements characteristics are as follows:

- Minimum-Maximum adsorber weight 5–600 g;
- Accuracy of the weighting unit ±0.1 g, time response, faster than 0.1 s;
- Accuracy of the temperature controllers of the circulating baths ±0.1 K;
- Accuracy of the pressure transducers ±0.1 mbar.

The system management and data acquisition is automatically implemented in LabVIEW® software.

Differently from the volumetric apparatus, the gravimetric one can allow investigating other parameters affecting the kinetic performance of a sorption TES, such as the water vapour transfer resistance inside complex heat exchanger geometries and the effect of heat transfer flow rate, which can give more practical information in the design phase of an innovative adsorber.

5.2 Characterization of Evaporators

Evaporators for adsorption storages need to be optimised to work under vacuum conditions with water as a refrigerant. However, theoretical studies and empirical or semi-empirical approaches for the description of the problem are still limited and therefore experimental systems for the description and evaluation of such components were developed.

At the RWTH University of Aachen, a suitable setup for the testing of tubes for evaporators working in sub-atmospheric conditions was realised. The experimental setup and testing procedure followed are described in [17]. The main aim of the experimental setup is to determine the overall heat transfer coefficient during evaporation as a function of the filling level of the refrigerant. It consists of a vacuum vessel, where the tubes to be tested are mounted horizontally. The refrigerant inside the chamber is thermally conditioned by means of a thermostatic bath. Instead, the pressure-induced evaporation is obtained by using a vacuum pump that removes water steam at a constant volume flow rate.

Sensors installed include:

- temperature sensors in the heat transfer fluid circuit;
- temperature sensor to measure water steam temperature in the evaporator chamber;
- pressure sensor in the evaporator chamber;
- flow meter in the HTF circuit.

The experimental procedure used is the following: at the beginning, the tubes are flooded with water and the vacuum pump is connected to the evaporator. No water is refilled during the measurement, thus, the filling level decreases continuously with time due to the evaporation of water. The measurement finally ends when the tubes fall dry and are no longer in contact with water.

The U-value for the process is defined as:

$$U = \frac{\dot{Q}}{A\Delta_{\ln}T} \quad (1)$$

where $\dot{Q}$ is the heat transferred by the HTF, A is the evaporation surface and $\Delta_{\ln}T$ is the logarithmic mean temperature difference between the HTF circuit and the water steam temperature. An example of the results obtained is shown in Fig. 16.

In [18], a test setup useful for the analysis of geometric and process parameter influencing sub-atmospheric evaporation under typical conditions for adsorption systems, and realised at Fraunhofer ISE, is presented. Differently from the previous case, it is suitable to test small-scale heat exchangers instead of tubes. It consists of two vacuum vessels: a “measurement chamber”, where the exchanger to be tested is installed, and a “secondary chamber”, that contains a permanently installed large tube-fin heat exchanger, which is partially flooded by refrigerant serving as a refrigerant reservoir. The heat exchangers are thermally conditioned by means of

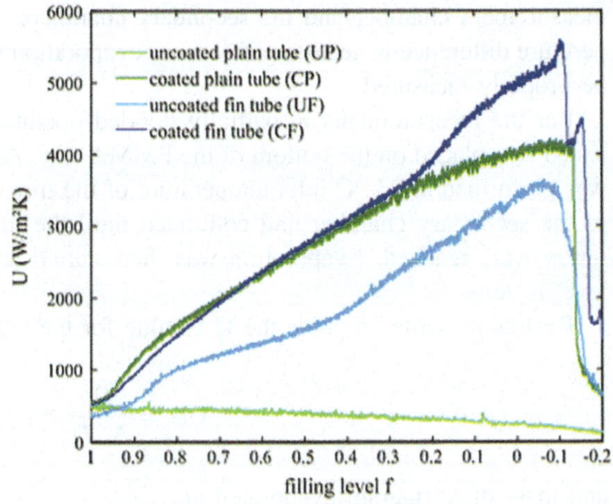

Fig. 16 Results obtained with the experimental setup for sub-atmospheric evaporators at RWTH Aachen [17]

circulation thermostats. The structure with a measurement chamber and a secondary chamber allows testing a heat exchanger both for condensation and evaporation processes.

Sensors installed include:

- Pt100 sensors to measure the temperature in the exchanger circuits and the ambient temperature;
- magnetic-inductive flow metres to measure the flow rate in the exchanger circuits;
- a capacitive pressure sensor and a piezoresistive pressure sensor to monitor pressure in the two chambers.

Two different operating modes are possible: evaporation of a thin refrigerant film from the heat exchanger surfaces and evaporation in a partially flooded mode with changing refrigerant filling level. Purpose of running two different modes was to cover the different application-related requirements of a power-focused (thin film) and a capacity-focused (partially flooded) evaporation. Prior to the installation in the chamber, the heat exchangers are coated to ensure comparability among the different specimens (by achieving identical surface conditions and a good wetting behaviour).

For the thin film evaporation measurement, the exchanger is placed in a transparent half-tube PMMA tray inside the measurement that has a bore on the bottom and is tilted, to allow draining of the refrigerant from the chamber. After careful evacuation, the fluid was pumped through the test specimen in the measurement chamber at 15 °C and the secondary chamber at 17 °C with closed connection valves. After 1 h (steady-state condition reached), the valves are open and the refrigerant is left to condense for 1 h. Excess refrigerant that cannot be hold on the surfaces drips off the heat exchanger into the tray and so is drained to the secondary chamber. After closing the valves, temperatures of 13 and 15°C were set in the measurement chamber and the secondary chambers, respectively. The small temperature difference is necessary to ensure evaporation dynamics is not too fast as to be properly measured.

For the measurements in partially flooded operation, the heat exchanger to be tested was placed on the bottom of the PMMA tray. After evacuation, condensation was performed at 23 °C inlet temperature of the measurement chamber and 19 °C of the secondary chamber and continued until the initial filling levels for evaporation were reached. Evaporation was then started and continued until the filling level is zero.

Results presented include the UA value for the heat exchanger:

$$\mathrm{UA} = \frac{\dot{Q}}{\Delta_{\ln} T} \tag{2}$$

and mass of refrigerant evaporated m_{evap}:

$$m_{\mathrm{evap}}(t) = \int_{t0}^{tf} \frac{\dot{Q}(\tau)}{\Delta h_{\mathrm{evap}}(\tau)} \mathrm{d}\tau \tag{3}$$

where $\Delta h_{\mathrm{evap}}(\tau)$ is the evaporation enthalpy.

Figure 17 shows an example of the results obtained for the tests under thin film evaporation, used for the evaluation of geometric parameters, and partially flooded mode, used to evaluate the effect of process parameters.

A third setup for the experimental testing of falling film evaporators, was developed at AEE Intec [19], whose schematics are presented in Fig. 18. The test rig consists of a vacuum vessel, containing the tube bundle to be tested, a condenser, and two reservoirs. One of the reservoirs serves as feeding source, while the

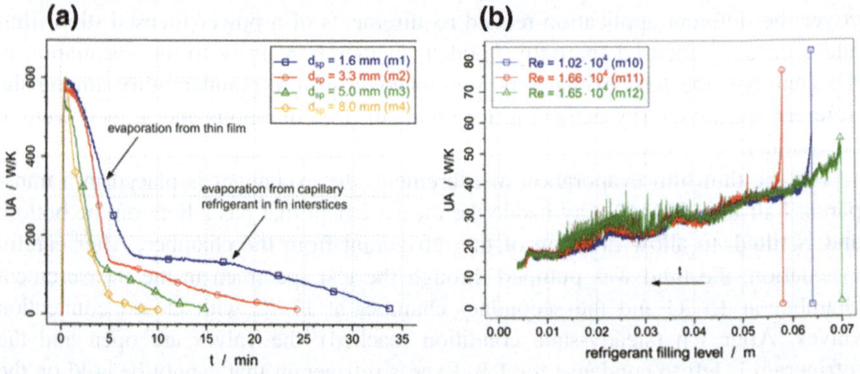

Fig. 17 Results of measurements in the testing rigs for sub-atmospheric evaporators at ISE. **a** UA for different fin densities during thin film evaporation experiments; **b** UA for a partially flooded test under different HTF regimes [18]

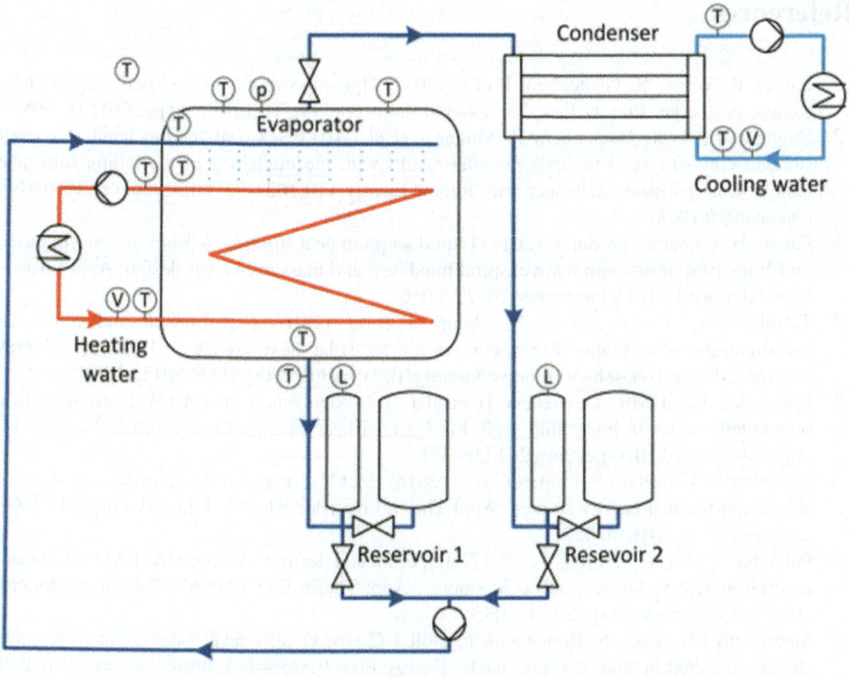

Fig. 18 Layout of the experimental setup at AEE Intec [19]

others collects surplus water. Apart from that, the test rig consists of three liquid circuits: the distilled water used for the evaporation (refrigerant), the water circuit for heating the tube bundle from the inside and the cooling water circuit for the condensation.

Sensors installed include:

- temperature sensors in the three hydraulic circuits;
- temperature sensors for measuring liquid and water steam temperatures in the evaporator;
- pressure sensor in the evaporator chamber;
- level sensor in the two reservoirs;
- ultrasonic flow meter installed between the reservoir and feeder in the refrigerant circuit for measuring the irrigation density;
- flow metres to measure the volumetric flow rate of the cooling water and heating water.

Results are presented in terms of the heat transfer coefficient. Indeed, from measurements, the overall *U*-value of the exchanger is derived, according to Eq. 1. Subsequently, the heat transfer coefficient inside the tube bundle is calculate according to Gnielinski correlation [19], and the mean outer heat transfer coefficient is used for performance evaluation.

References

1. Fumey B, Weber R, Gantenbein P et al (2015) Operation results of a closed sorption heat storage prototype. Energy Proc 73:324–330. https://doi.org/10.1016/j.egypro.2015.07.698
2. Daguenet-Frick X, Gantenbein P, Müller J et al (2017) Seasonal thermochemical energy storage: comparison of the experimental results with the modelling of the falling film tube bundle heat and mass exchanger unit. Renew Energy 110:162–173. https://doi.org/10.1016/j.renene.2016.10.005
3. Fumey B, Weber R, Baldini L (2017) Liquid sorption heat storage—a proof of concept based on lab measurements with a novel spiral fined heat and mass exchanger design. Appl Energy. https://doi.org/10.1016/j.apenergy.2017.05.056
4. Tatsidjodoung P, Le Pierrès N, Heintz J et al (2016) Experimental and numerical investigations of a zeolite 13X/water reactor for solar heat storage in buildings. Energy Convers Manag 108:488–500. https://doi.org/10.1016/J.ENCONMAN.2015.11.011
5. Gaeini M, Javed MR, Ouwerkerk H et al (2017) Realization of a 4 kW thermochemical segmented reactor in household scale for seasonal heat storage. Energy Proc 135:105–114. https://doi.org/10.1016/j.egypro.2017.09.491
6. Schreiber H, Lanzerath F, Reinert C et al (2016) Heat lost or stored: Experimental analysis of adsorption thermal energy storage. Appl Therm Eng 106:981–991. https://doi.org/10.1016/j.applthermaleng.2016.06.058
7. Palomba V, Vasta S, Freni A (2017) Experimental testing of AQSOA FAM Z02/water adsorption system for heat and cold storage. Appl Therm Eng 124:967–974. https://doi.org/10.1016/j.applthermaleng.2017.06.085
8. Melograno PN, Vasta S, Boudehenn F, Döll J (2016) Quality level assessment of sorption chillers installed in solar cooling plants. Energy Proc 91:356–365. https://doi.org/10.1016/J.EGYPRO.2016.06.283
9. Aristov YI, Dawoud B, Glaznev IS, Elyas A (2008) A new methodology of studying the dynamics of water sorption/desorption under real operating conditions of adsorption heat pumps: Experiment. Int J Heat Mass Transf 51:4966–4972. https://doi.org/10.1016/J.IJHEATMASSTRANSFER.2007.10.042
10. Grekova AD, Gordeeva LG, Aristov YI (2017) Composite "LiCl/vermiculite" as advanced water sorbent for thermal energy storage. Appl Therm Eng 124:1401–1408. https://doi.org/10.1016/J.APPLTHERMALENG.2017.06.122
11. Grekova AD, Gordeeva LG, Lu Z et al (2018) Composite "LiCl/MWCNT" as advanced water sorbent for thermal energy storage: sorption dynamics. Sol Energy Mater Sol Cells 176:273–279. https://doi.org/10.1016/J.SOLMAT.2017.12.011
12. Sapienza A, Santamaria S, Frazzica A et al (2014) Dynamic study of adsorbers by a new gravimetric version of the large temperature jump method. Appl Energy 113:1244–1251. https://doi.org/10.1016/J.APENERGY.2013.09.005
13. Brancato V, Gordeeva L, Sapienza A et al (2016) Dynamics study of ethanol adsorption on microporous activated carbon for adsorptive cooling applications. Appl Therm Eng 105:28–38. https://doi.org/10.1016/j.applthermaleng.2016.05.148
14. Gordeeva L, Frazzica A, Sapienza A et al (2014) Adsorption cooling utilizing the "LiBr/silica–ethanol" working pair: dynamic optimization of the adsorber/heat exchanger unit. Energy 75:390–399. https://doi.org/10.1016/j.energy.2014.07.088
15. Graf S, Lanzerath F, Sapienza A et al (2016) Prediction of SCP and COP for adsorption heat pumps and chillers by combining the large-temperature-jump method and dynamic modeling. Appl Therm Eng 98:900–909. https://doi.org/10.1016/j.applthermaleng.2015.12.002
16. Santamaria S, Sapienza A, Frazzica A et al (2014) Water adsorption dynamics on representative pieces of real adsorbers for adsorptive chillers. Appl Energy 134:11–19. https://doi.org/10.1016/J.APENERGY.2014.07.053

17. Lanzerath F, Seiler J, Erdogan M et al (2016) The impact of filling level resolved: capillary-assisted evaporation of water for adsorption heat pumps. Appl Therm Eng. https://doi.org/10.1016/j.applthermaleng.2016.03.052
18. Volmer R, Eckert J, Füldner G, Schnabel L (2017) Evaporator development for adsorption heat transformation devices—influencing factors on non-stationary evaporation with tube-fin heat exchangers at sub-atmospheric pressure. Renew Energy 110:141–153. https://doi.org/10.1016/J.RENENE.2016.08.030
19. Dang BN, Van Helden W, Luke A (2017) Investigation of water evaporation for closed sorption storage systems. Energy Proc 135:504–512. https://doi.org/10.1016/j.egypro.2017.09.493

Definition of Performance Indicators for Thermal Energy Storage

Valeria Palomba, Jaume Gasia, Joaquim Romaní, Andrea Frazzica and Luisa F. Cabeza

Abstract With the aim of standardizing the evaluation of thermal storage systems/tanks, this chapter assesses and compares the different performance indicators that can be found in the literature and tries to recommend those which enable a better comparison.

1 Introduction

Thermal energy storage (TES) technologies are already a reality. However, despite the fact that research on storage materials has been deeply addressed for decades, the maturity of alternative systems to sensible heat TES is still limited [1]. While latent heat TES applications have been made commercially available recently, chemical and sorption systems for storage purposes are still not commercially available, despite an increasing number of prototypes being developed at laboratory scale [2]. In view of advancement towards proper designing and commercialization of TES systems, a standardized TES analysis procedure is needed. An example of the importance of a standardization procedure is clear when considering the work presented in the previous chapters. In these, several experimental activities reported in literature were assessed, and it could be noticed that different parameters and definitions were adopted, which complicate the interpretation of the experimental data, making them hardly comparable. One way to address this issue and make a step forward towards the standardized analysis procedure for TES systems comparison is the definition of shared performance indicators.

V. Palomba (✉) · A. Frazzica
CNR-ITAE, Istituto di Tecnologie Avanzate per l'Energia "Nicola Giordano", Via Salita S. Lucia Sopra Contesse 5, 98126 Santa Lucia, Messina, Italy
e-mail: valeria.palomba@itae.cnr.it

J. Gasia · J. Romaní · L. F. Cabeza
GREiA Research Group, INSPIRES Research Centre, Universitat de Lleida, Pere de Cabrera s/n, 25001 Lleida, Spain

A. Frazzica and L. F. Cabeza (eds.), *Recent Advancements in Materials and Systems for Thermal Energy Storage*, Green Energy and Technology,
https://doi.org/10.1007/978-3-319-96640-3_13

In the present chapter, an attempt to discuss the relevant indicators for TES evaluation and comparison is reported. This approach follows the methodology developed within the framework of IEA-ECES Annex 30 'Thermal Energy Storage for Cost-Effective Energy Management and CO_2 Mitigation' [3]. The aim is to propose a set of performance indicators that covers different layers in the TES technology, such as technical and economical, and that provide the generalities for their definition within the various technologies. Subsequently, a set of technical parameters, chosen among the presented ones, were applied to two case studies at different development levels with the final goal of demonstrating the applicability of the methodology proposed. Indeed, the analysed systems are prototypes and therefore a complete and systematic assessment of technical and economical parameters is not possible. Nonetheless, since no previous attempts at applying common indicators to different TES technologies are reported in the literature, the results here reported are still valuable and demonstrate the applicability of the methodology.

2 Performance Indicators

Among the parameters, or indicators, which are useful to characterize a TES in a diverse range of aspects, the most relevant are known as performance indicators (PI). The main features that a performance indicator should possess for a proper evaluation and comparison are:

- **Uniqueness**: It should have a single definition in order to not generate confusion or misunderstanding.
- **Applicability**: It should not be limited to specific cases, sizes and processes but to a wide range of them.
- **Measurability**: It should be based on accessible quantitative data or objective qualitative criterions.
- **Relevance**: It should be meaningful to the purposes of the TES and specific to the characteristics of its processes and components.

Among the different performance indicators defined for the characterization of a TES system or for the comparison of different TES systems, there are a set of them which are relevant for the final application and depend on the stakeholders involved. They are known as Key Performance Indicators (KPI). KPIs strongly depend on the specific applications considered and the stakeholders involved. For example, the main requirements for a concentrated solar power (CSP) plant or a district heating network are dissimilar and therefore the performance indicators relevant in the different cases are equally dissimilar: e.g., in case of CSP, cycle length is a constraint that is not of such relevance for a district heating network. An even clearer example is obtained when comparing TES for mobile applications (e.g. automotive) with stationary ones (e.g. residential sector): in mobile applications, storage weight is a key parameter, which is not the case of residential storages.

3 Identification of Performance Indicators for TES Systems Evaluation

The analysis of a TES system requires a multi-level evaluation, covering not only the technical aspects related to the design and installation of the system, but also economic and life cycle aspects, following the complex demands that a TES system should fulfil, among others. An overview of some of the most representative performance indicators for TES systems is given in Fig. 1. In the following sections, a brief explanation of their meaning and definition is given.

3.1 Technical Performance Indicators

Technical performance indicators refer to parameters related to the design, operation and control of a TES system. Within this group, four different subgroups are proposed: structure, energy, dynamics and development.

3.1.1 Structure Performance Indicators

Structure performance indicators are defined with the aim of typifying the physical characteristics of the system. Useful structure performance indicators are the mass and volume of the storage material on the total mass or volume of the TES module.

The **storage material mass ratio** is calculated as the mass of TES material over the mass of the TES module:

$$m_{\text{mat-sys}} = \frac{m_{\text{TES.material}}}{m_{\text{TES.system}}} \left[\frac{\text{kg}}{\text{kg}}\right] \tag{1}$$

The **storage material volume ratio**, also known as packing factor, is the volume occupied by the storage material in front of the volume of the TES module:

$$V_{\text{mat-sys}} = \frac{V_{\text{TES.material}}}{V_{\text{TES.system}}} \left[\frac{\text{m}^3}{\text{m}^3}\right] \tag{2}$$

Notice that in the definition of the mass or volume of the storage, the selection of the TES module boundary is essential to decide which components have to be included or excluded from the calculation. In the current indicators, the mass or

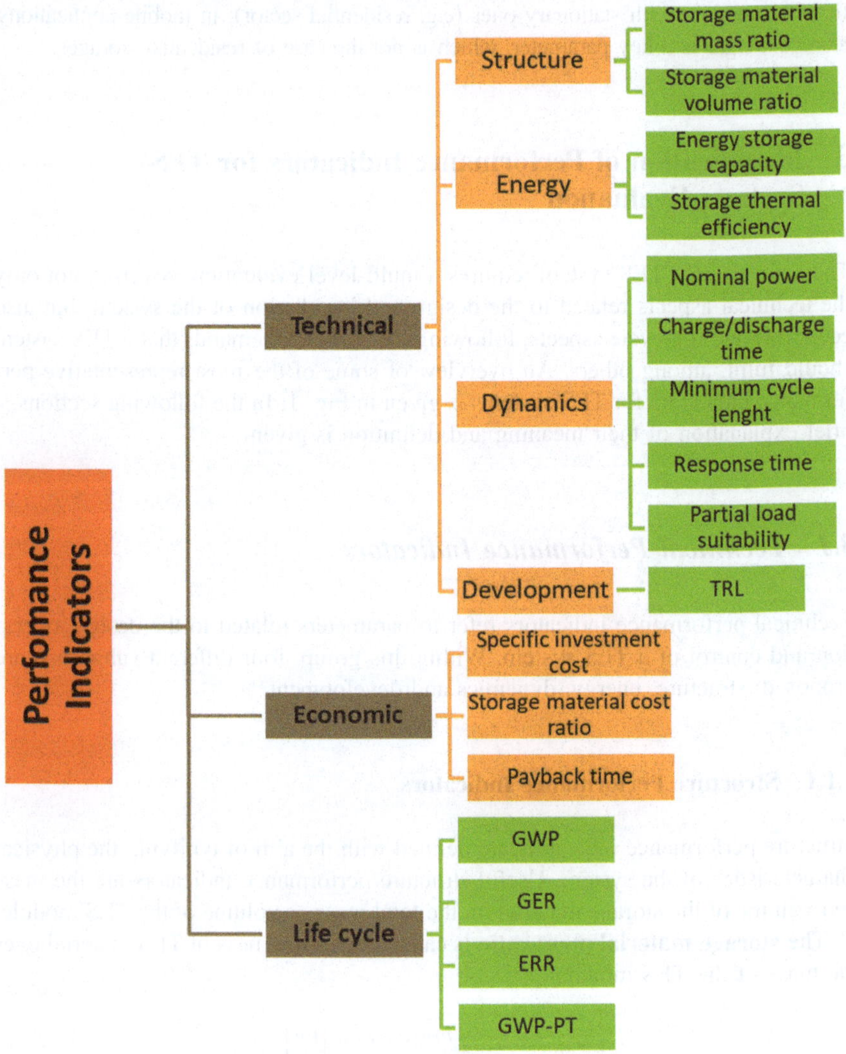

Fig. 1 Performance indicators selected and analysed in the present chapter

volume of the TES tank are considered at the 'container' level, as shown in Fig. 2. Hence, for sensible and latent TES the heat exchanger and the casing of the material are evaluated, without any consideration for the insulation or the external piping and manifolds. In the case of the sorption TES, the difficulty lies in the definition of the tank boundary since not only a reactor but also an evaporator/condenser is needed to make the system work. The technology, structure and location of the evaporator/condenser are completely independent from that of the ad/absorber and thus, the boundary is considered around the reactor, without any insulation.

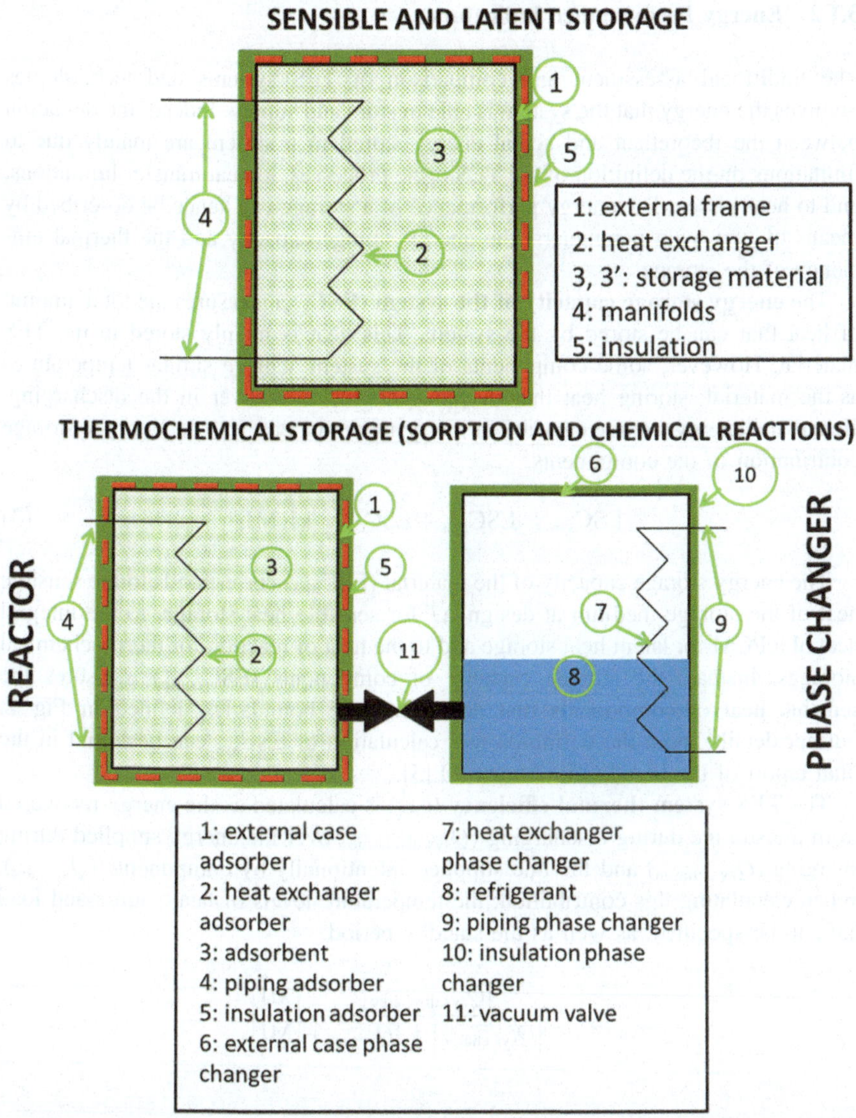

Fig. 2 Boundaries of the storage for the different technologies

Furthermore, it is worth noticing that, even if the two indicators are directly linked via the density of storage material and storage module, the distinct definition of both of them is important in view of KPIs for different applications. For example, the storage material mass ratio can be more important than storage material volume ratio in some mobile applications where higher mass imply higher primary energy consumption (e.g. fuel), while volume might not represent a key constraint.

3.1.2 Energy Performance Indicators

The traditional assessment and comparison of TES systems and technologies involves the energy that the system is able to store and supply. Indeed, the deviation between the theoretical and actual energy stored in a system are mainly due to limitations on the definition of the TES cycle processes, to heat transfer limitations, and to heat losses. The energy performance of a storage can hence be described by means of two main parameters: the energy storage capacity and the thermal efficiency of the storage.

The **energy storage capacity of the system (ESC_{sys})** measures the total amount of heat that can be stored by the system. This heat is mainly stored in the TES material. However, some components of the system achieve similar temperatures as the material, storing heat that may be possible to recover in the discharging. Hence, the energy storage capacity of the system must take into account the storage contribution of the components.

$$ESC_{sys} = ESC_{mat} + ESC_{comp}\ [kJ, MJ] \tag{3}$$

The energy storage capacity of the material (ESC_{mat}) corresponds to the sensible heat of the storage medium at design ΔT for sensible heat storage, to the integral heat of a PCM for latent heat storage and to the heat of reaction for thermochemical storages. Instead, the storage capacity of components (ESC_{comp}) measures the sensible heat of components that are within the boundaries defined in Fig. 2. Further details about the definition and calculation of ESC_{sys} can be found in the final report of the IEA-ECES Annex 30 [3].

The **TES system thermal efficiency (ε_{sys})** is calculated as the energy recovered from the storage during discharging ($Q_{sys.discharge}$) over the energy supplied during charging ($Q_{sys.charge}$) and the one supplied intentionally by components ($Q_{sys.aux}$). When calculating this contribution, the temperature levels of heat source and load have to be specified, as well as the standby period.

$$\varepsilon_{sys} = \frac{|Q_{sys.discharge}|}{|Q_{sys.charge}| + |Q_{sys.aux}|} \left[\frac{MJ}{MJ}\right] \tag{4}$$

3.1.3 Dynamics Performance Indicators

Dynamics performance indicators are parameters which are related to the behaviour of the TES system during their operation. Considerations on the energy that a TES can store and supply cannot preclude from evaluations on the effect of dynamics. Therefore, the dynamics performance indicators of a TES system are described by the following parameters: nominal power, charge and discharge time, or response time.

The **nominal power ($Q_{nom.sys.discharge}$), *measured in* kW**, is the design thermal power supplied by the storage during discharge. However, the nominal power for charge might also represent a relevant parameter and can also be indicated as $Q_{nom.sys.charge}$.

The **charge/discharge time** (t_{charge}, $t_{discharge}$), measured in minutes, hours or days, represents the time needed to complete the charge or discharge process.

The definition of charge and discharge time is linked to the **minimum cycle length** ($MinCy_{sys}$), which represents the shortest period required for completely charging and discharging the system at nominal conditions, without considering any storage or standby process. It can be calculated as:

$$\mathrm{MinCy_{sys}} = \frac{\mathrm{ESC_{sys}}}{Q_{\mathrm{nom.sys.charge}}} + \frac{\mathrm{ESC_{sys}}}{Q_{\mathrm{nom.sys.discharge}}} [\mathrm{min}, h, d] \quad (5)$$

The response time ($ReTi_{sys}$) is the time passing between the connection of the storage to the load until nominal power is reached. As for the previous performance indicators, separate values for charge and discharge could be indicated, even though the ones related to discharge are the most relevant.

Finally, the **partial load suitability (PL_{sys})** is a qualitative parameter that indicates whether partial charge or discharge is possible. Possible values for such a parameter are not suitable, partially suitable and suitable. The criterions for defining the part load suitability are the capability of the TES system for stopping the charging or discharging process, or the capability to switch swiftly from one to the other.

3.1.4 Development

***Development of technology can be described by means of* technology readiness level** (TRL). This is a parameter which measures the technology maturity and gives information on the state of development of the system under analysis [4]. It establishes a qualitative approach to the development of a technology by defining nine categories that consider the level at which the concept has been tested or developed, going from testing of the basic principles (TRL 1) to the actual system proven in operational environment (TRL 9).

3.2 Economic Performance Indicators

The economic performance indicators can be applied for the assessment of systems that are in pre-commercial or commercial stage. Relevant indicators of the economic performance of the system include the specific investment cost and the payback time.

Two economic PIs commonly used in different fields are CAPEX and OPEX. **CAPEX** refers to capital expenses to acquire the storage module, while **OPEX** refers to the operating expenses to run the storage module.

In addition, it is possible to identify some more specific indicators, such as the specific investment cost and the storage material cost ratio.

The **specific investment cost (IC_{sys})** is the overall cost of the system normalized according to its energy storage capacity. Hence, it includes the cost of the storage material (C_{mat}), the cost of the components (C_{comp}) and the different accessory expenses (C_{aux}) (e.g. manufacturing):

$$\mathrm{IC}_{\mathrm{sys}} = \frac{C_{\mathrm{mat}} + C_{\mathrm{comp}} + C_{\mathrm{aux}}}{\mathrm{ESC}_{\mathrm{sys}}} \left[\frac{€}{\mathrm{MJ}}\right] \tag{6}$$

The ratio of the overall investment cost that is due to the storage material represents the **storage material cost ratio ($C_{mat\text{-}sys}$)**:

$$C_{\mathrm{mat-sys}} = \frac{C_{\mathrm{mat}}}{C_{\mathrm{mat}} + C_{\mathrm{comp}} + C_{\mathrm{aux}}} \left[\frac{€}{€}\right] \tag{7}$$

Finally, the **payback time** PT_{sys} represents the time needed to get the investment back. It can be calculated as a simple or discounted payback time.

3.3 Life Cycle Performance Indicators

Recently, the approach in environmental analyses has shifted towards an evaluation of the impact of a technology throughout its entire life cycle [5]. This ‘cradle to grave’ approach is referred as life cycle analysis (LCA), and it has become especially relevant for stakeholders such regulatory agencies.

Among the output of LCA methodologies, useful performance indicators are the **global energy requirement** (GER_{sys}) and the **global warming potential** (GWP_{sys}).

GER_{sys} represents an indicator of the overall energy consumption, during the whole life cycle, of a functional unit of the storage module. It includes the energy for the production as well as the one needed to produce the energy vectors used in production processes and transport, as well as the feedstock energy of raw materials. GWP_{sys} represents the carbon footprint of the storage module.

4 Example of Performance Indicators Calculation for a Latent Heat Storage System

4.1 *Description of the System*

The applicability of the performance indicators is further studied in a latent heat TES system at pilot plant level at the University of Lleida [6]. This system consists of a storage tank based on the shell-and-tube heat exchanger concept and contains high density polyethylene (HDPE) as phase change material (PCM). The module considered in this section is a prototype of this technology used for research purposes, hence it was downscaled compared to the real application. Consequently, it was a useful case study for benchmarking the structure, energy, and dynamics performance indicators presented in the previous section. However, being a prototype for research, the system cannot provide extensive data for evaluating economic and life cycle PI.

The main characteristics of the latent TES are reported in Table 1 and Fig. 3.

The test used to calculate the dynamic and energy performance indicators is explained in detail in Gasia et al. [6]. The main features and results of this test are summarized in Table 2.

4.2 *Results*

4.2.1 Structure Performance Indicators

The latent heat TES system was simple to study in terms of structure performance indicators. Being a single shell-and-tube module it is simple to find the boundaries of the container, as only the insulation was excluded from the calculation in this case. The result of the structure parameters is presented in Table 3.

Table 1 Main features of the latent heat TES system used at the University of Lleida

Dimensions of the module (no insulation)	0.53 × 0.27 × 1.27 m
Number of tubes	49 units
Heat transfer surface	6.55 m^2
Heat transfer fluid	Sylicon fluid Syltherm 800
Storage material	High density polyethylene (HDPE)
Phase change range	124–134 °C (melting) 126–114 °C (solidification)
System operation temperature range	105–155 °C

Fig. 3 Overview of the latent heat TES system used at the University of Lleida

Table 2 Test data for charge, storage, and discharge cycle of HDPE latent storage

Process	HTF mass flow rate (kg s^{-1})	HTF inlet temperature (°C)	PCM average initial temperature (°C)
Charge	0.5	155	104
Discharge	0.5	105	150

Table 3 Structure performance indicators results for HDPE latent TES

Parameter	Value	Units	Parameter	Value	Units
V_{mat}	0.133	m^3	m_{mat}	99.5	kg
V_{TES}	0.154	m^3	m_{mat}	176.2	kg
$V_{mat\text{-}TES}$	86.4	%	$m_{mat\text{-}TES}$	56.46	%

4.2.2 Energy Performance Indicators

As a latent heat TES system, the HDPE module mainly stores heat through the melting enthalpy of the material. Moreover, in the present case study the TES material is in direct contact with the tubes, the baffles, and separators of the TES system. Thus, these components are assumed to stay at the same temperature than the material and, as a consequence, they are accounted in the total energy storage capacity (ESC) of the system. Furthermore, the system operates under a temperature difference between charge and discharge of 50 K, which extends beyond the phase change range, and thus the material also stores sensible heat. The energy storage

Table 4 Energy storage capacity of the HDPE latent TES

Parameter		Value	Units
Mass of the metal components which contribute to the energy storage capacity of the system	m_{comp}	121.6	kg
Specific heat of the components	$C_{p\text{-}comp}$	0.53	kJ kg K^{-1}
Temperature gradient of the system	ΔT_{sys}	50.0	K
Specific heat of the TES material at solid state	$C_{p.mat.solid}$	2.6	kJ/kg K
Melting enthalpy of the TES material	$\Delta H_{melting}$	138	kJ/kg K
Specific heat of the TES material at liquid state	$C_{p.mat.liquid}$	2.7	kJ/kg K
Mass of the TES material	m_{mat}	99.5	kg
Energy storage capacity of the TES material	ESC_{mat}	24.29	MJ
Energy storage capacity of the components	ESC_{comp}	3.22	MJ
Energy storage capacity of the system	ESC_{sys}	27.51	MJ

capacity of the latent TES studied is presented in Table 4, where all the variables and parameters used in the calculation are shown.

Then, the thermal efficiency the system is evaluated with absorbed and released heat during the charging and discharging processes. The latent storage system does not have any components that generate auxiliary heat, hence the efficiency was simply a ratio between the charging and discharging heat, as follows:

$$\varepsilon_{sys} = \frac{|Q_{sys.discharge}|}{|Q_{sys.charge}| + |Q_{sys.aux}|} = \frac{8.75}{9.21 + 0} = 0.95$$

4.2.3 Dynamics Performance Indicators

The input temperatures of the HTF to the storage module are 155 and 105 °C for charging and discharging processes, respectively. Accounting for the temperature difference of the HTF with the HDPE, the mass flow, and the characteristics of the tanks, which can be assimilated to a shell-and-tubes heat exchanger, the nominal power is:

$$Q_{nom.sys.charge} = 3.5\,\text{kW}$$

The other dynamics performance indicators are calculated according to the nominal power and the storage capacity presented in Table 5.

Table 5 Dynamics PIs for the latent HDPE TES developed at UdL

Test	τ_{charge}	$\tau_{discharge}$	$MinCy_{sys}$	$RETi_{sys.discharge}$
	min	min	h	min
1	332	190	4.37	60

The values for τ_{charge} and $\tau_{discharge}$ in Table 5 correspond to the time needed to reach 90% of the storage capacity. Indeed, especially during the charging process, the effect of heat losses becomes relevant when going from 90 to 100% of storage capacity. A complete and detailed discussion of the issue is given in [6].

Moreover, according to the test carried out in [6], the latent heat storage with HDPE is considered as suitable for partial load operation, although this affected the heat transfer rate of the discharging process.

4.2.4 Development

The latent heat TES system was tested in a relevant environment, where its inputs and outputs simulated the behaviour of an industry. Therefore, its development level according to the definitions reported in [4] has to be considered as TRL 5.

5 Example of Performance Indicators Calculation for a Physical Sorption Storage System

5.1 Description of the System

In order to demonstrate the applicability of the selected performance indicators also to thermochemical storages, an adsorption TES developed at ITAE and extensively described in [7] is presented as a case study. It is however worth stressing that the system chosen is a prototype of the technology and therefore, it is not possible to calculate all the parameters defined above, especially the economic and life cycle PIs. Hence, the focus of this section is on the evaluation of technical performance indicators that represent the basis for the design analysis, comparison, and optimization.

The main characteristics of the selected storage are reported in Table 6, while a picture of the system is shown in Fig. 4.

The calculation of the energy and dynamics performance indicators is carried out taking into account that an adsorption storage can be employed for heat or cold storage purposes. Consequently, it is possible to calculate the relevant performance indicators under these two operating modes. Indeed, in [7] three cases are considered: cold storage, short-term heat storage, and long-term heat storage. For the sake of simplicity only cold storage and long-term (seasonal) heat storage are considered in the present calculations. The results consider six different tests, mainly differing for the boundary conditions in terms of temperature during both charge and discharge. The list of tests and their conditions is presented in Table 7.

Table 6 Main features of the adsorption TES developed at CNR-ITAE

Dimensions of the module (no insulation)	250 × 553 × 774 mm
Adsorber heat exchanger	1 aluminium flat-tube heat exchanger
heat transfer surface adsorber	1.75 m^2
Adsorbent material	4.3 kg AQSOA FAM-Z02
Adsorber mass	19 kg
Adsorber volume	3.5 dm^3
Overall dimensions of the phase changer chamber	304 × 177 × 298 mm
Phase changer heat exchangers	4 × copper/SS tube-and-fin heat exchangers
Heat transfer surface phase changer	5.32 m^2
Heat transfer fluid	Water
Refrigerant	Water
Amount of refrigerant	4 L

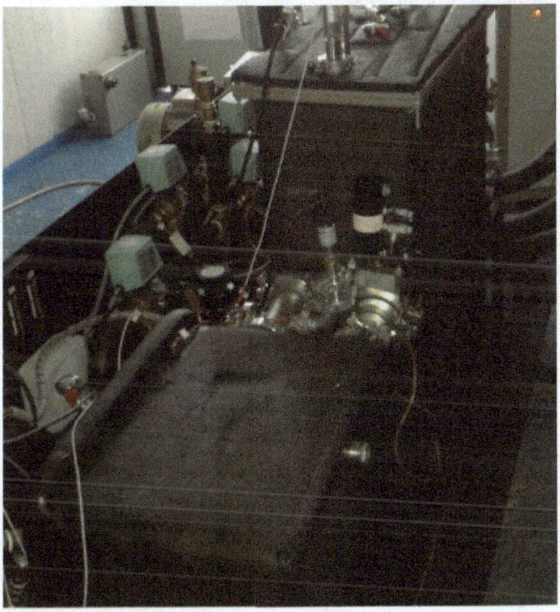

Fig. 4 The adsorption TES system developed at CNR-ITAE

5.2 *Results*

5.2.1 Structure Performance Indicators

The calculation of the structural features of the adsorption storage according to the boundaries of Fig. 2 presents a significant issue: the chamber in which the

Table 7 Test numbers and testing conditions for the calculation of the energy and dynamics performance indicators

Test	Charge				Discharge			
	T_{des}	T_{cond}	Flow adsorber	Flow phase changer	$T_{adsorption}$	$T_{evaporator}$	Flow adsorber	Flow phase changer
	°C	°C	kg/min	kg/min	°C	°C	kg/min	kg/min
1	92	14	11	15	35	9	12	15
2	82	29	13	12	36	11	12	14
3	75	30	12	12	35	11	11	14
4	92	39	12	13	40	6	12	13
5	92	41	12	12	41	15	12	14
6	82	15	13	15	30	6	11	14

Table 8 Structure performance indicators results for the adsorption TES

Parameter	Value	Units	Parameter	Value	Units
V_{mat}	0.035	m^3	m_{mat}	4.3	kg
V_{TES}	0.10	m^3	m_{TES}	18.7	kg
$V_{mat\text{-}TES}$	31	%	$m_{mat\text{-}TES}$	23	%

adsorbent material and the heat exchange are inserted is significantly larger than the developed unit (material + heat exchanger), since it is designed for the testing of different systems at different scales. As a consequence, the structural performance indicators would be extremely unfavourable: it was then decided to refer the calculation to the volume and weight of the adsorber, meant as the material and the heat exchanger in which it is contained. The results are reported in Table 8.

5.2.2 Energy Performance Indicators

The results of the energy PIs, calculated for seasonal heat storage and cold storage conditions and the selected tests, are presented in Table 9.

5.2.3 Dynamics Performance Indicators

The calculation of dynamics performance indicators was carried out for the same six tests listed in Table 7. All the parameters are presented for charge and discharge.

The nominal powers are:

$$Q_{nom.sys.charge} = 1\,\text{kW}$$

Table 9 Energy performance indicators for the adsorption TES developed at CNR-ITAE under different testing conditions

Test	ESC_{sys} CS	ESC_{sys} SHS	ε_{sys} CS	ε_{sys} SHS
	MJ	MJ	–	–
1	1.05	2.56	0.38	0.56
2	0.97	2.03	0.36	0.41
3	0.67	1.53	0.29	0.38
4	0.76	1.59	0.33	0.35
5	1.21	2.35	0.42	0.39
6	0.86	2.18	0.30	0.45

Table 10 Dynamics performance indicators for the adsorption TES system developed at CNR-ITAE

Test	t_{charge}	$t_{discharge}$	$MinCy_{sys}$ CS	$RETi_{sys.charge}$	$RETi_{sys.discharge}$
	min	min	h	s	s
1	109	321	1.26	5	8
2	325	342	1.17	10	8
3	421	509	0.80	17	8
4	430	428	0.91	5	13
5	408	465	1.45	5	13
6	279	392	1.03	10	5

$$Q_{nom.sys.discharge} = \begin{cases} 0.5\,\text{kW seasonal heat storage} \\ 0.3\,\text{kW cold storage} \end{cases}$$

The adsorption TES is expected to be classified as suitable regarding partial load suitability PI, but further tests should be carried out to experimentally prove the effect of partial load operating conditions on this type of storage. The other dynamics performance indicators calculated are reported in Table 10.

5.2.4 Development

According to the definitions reported in [4], the system is at TRL 4, corresponding to 'technology validated in lab'.

6 Discussion

The results presented in the previous sections demonstrate that it was possible to apply the definitions of the technical performance indicators proposed to two prototypal systems, even if in a lab-scale. The calculation allowed identifying some critical issues. The clear definition of the boundary of the system is of the uttermost importance, since it significantly affects the final results. However, the definition of the different PIs per se proved to be clear and complies with the requirements of uniqueness, measurability and relevance. Indeed, it is possible to state that the complete overview of the different technical PIs exhaustively delineates the main features of the proposed systems even to users who were not involved in its design in the first place.

References

1. Cabeza LF (n.d.) Advances in thermal energy storage systems : methods and applications
2. Scapino L, Zondag HA, Van Bael J, Diriken J, Rindt CCM (2017) Sorption heat storage for long-term low-temperature applications: a review on the advancements at material and prototype scale. Appl Energy 190:920–948. https://doi.org/10.1016/j.apenergy.2016.12.148
3. ECES Annex 30 (n.d.) Thermal energy storage for cost-effective energy management. http://www.eces-a30.org/. Accessed 22 March 2018
4. European Commission (n.d.) HORIZON 2020—Work Programme 2018–2020 General Annex G. https://ec.europa.eu/research/participants/data/ref/h2020/other/wp/2018-2020/annexes/h2020-wp1820-annex-g-trl_en.pdf. Accessed 22 March 2018
5. IPCC (2013) Climate change 2013: The physical science basis. In: Contribution of working group I to the fifth assessment report of the intergovernmental panel on climate change. Stocker TF, Qin D, Plattner, G-K, Tignor M, Allen, SK, Boschung J, Nauels A, Xia Y, Bex V, Midgley PM (eds) Cambridge University Press, Cambridge, United Kingdom and New York, NY, USA, 2014. https://doi.org/10.1017/cbo9781107415324
6. Gasia J, de Gracia A, Peiró G, Arena S, Cau G, Cabeza LF (2018) Use of partial load operating conditions for latent thermal energy storage management. Appl Energy 216:234–242. https://doi.org/10.1016/J.APENERGY.2018.02.061
7. Palomba V, Vasta S, Freni A (2017) Experimental testing of AQSOA FAM Z02/water adsorption system for heat and cold storage. Appl Therm Eng 124:967–974. https://doi.org/10.1016/J.APPLTHERMALENG.2017.06.085

The manufacturer's authorised representative in the EU is Springer Nature Customer Service Centre GmbH, Europaplatz 3, 69115 Heidelberg, Germany. If you have any concerns regarding our products, please contact ProductSafety@springernature.com

Printed and bound by CPI Group (UK) Ltd, Croydon, CR0 4YY

30/09/2026

02225643-0007